# The Amusement Park

Cyclone, Revolution, Corkscrew; Luna Park, Pleasure Beach, Dreamland – names and places instantly familiar to rollercoaster and amusement park enthusiasts. But what first gave rise to the concept and nomenclature of the amusement park; how did amusement parks develop in Britain and elsewhere, and what fate awaits historic amusement parks and their rides today?

This thought-provoking and timely book brings together leading writers from a variety of disciplines to explore the social history and cultural heritage of the amusement park. Rooted in the British experience but informed by extensive international coverage, it provides a thematic, comparative exploration of the origins, development, decline and significance of the amusement park. The rich set of case studies presented comment on the interrelationships between history, culture and heritage, challenging traditional academic boundaries while offering important contributions to policy-making and regeneration initiatives.

The book provides new insights into a neglected aspect of popular culture and will be a valuable resource to students and scholars of history, heritage, tourism, leisure, technology and design.

**Jason Wood** is Director of Heritage Consultancy Services, Lancaster, UK, and former Professor of Cultural Heritage at Leeds Metropolitan University, UK.

# Heritage, Culture and Identity

*Series editor: Brian Graham, University of Ulster, UK*

This series explores all notions of heritage – including social and cultural heritage, the meanings of place and identity, multiculturalism, management and planning, tourism, conservation, and the built environment – at all scales from the global to the local. Although primarily geographical in orientation, it is open to other disciplines such as anthropology, history, cultural studies, planning, tourism, architecture/conservation, and local governance and cultural economics.

For a full list of titles in this series, please visit
www.routledge.com/Heritage-Culture-and-Identity/book-series/ASHSER-1231

**Cultures of Race and Ethnicity at the Museum**
Exploring Post-imperial Heritage Practices
*Divya P. Tolia-Kelly*

**Defining National Heritage**
The National Trust from Open Spaces to Popular Culture
*Leslie G. Cintron*

**World Heritage Sites and Tourism**
Global and Local Relations
*Edited by Laurent Bourdeau, Maria Gravari-Barbas and Mike Robinson*

**Heritage, Conservation and Communities**
Engagement, Participation and Capacity Building
*Edited by Gill Chitty*

**The Amusement Park**
History, Culture and the Heritage of Pleasure
*Edited by Jason Wood*

# The Amusement Park

History, Culture and the
Heritage of Pleasure

**Edited by Jason Wood**

Routledge
Taylor & Francis Group

LONDON AND NEW YORK

First published 2017
by Routledge
2 Park Square, Milton Park, Abingdon, Oxon OX14 4RN

and by Routledge
711 Third Avenue, New York, NY 10017

*Routledge is an imprint of the Taylor & Francis Group, an informa business*

*British Library Cataloguing in Publication Data*
A catalogue record for this book is available from the British Library

*Library of Congress Cataloging in Publication Data*
Names: Wood, Jason, 1959- editor.
Title: The amusement park : history, culture and the heritage of
    pleasure / edited by Jason Wood.
Description: New York : Routledge, 2017. | Series: Heritage, Culture
    and Identity
Identifiers: LCCN 2016037514| ISBN 9781472423726 (Hardback) |
    ISBN 9781315612409 (eBook)
Subjects: LCSH: Amusement parks—History. | Amusement parks—
    Social aspects.
Classification: LCC GV1851.A35 A555 2017 | DDC 791.06/8—dc23
LC record available at https://lccn.loc.gov/2016037514

ISBN: 978-1-4724-2372-6 (hbk) ISBN 9780367022709 (pbk)
ISBN: 978-1-315-61240-9 (ebk)

Typeset in Times New Roman
by Swales & Willis Ltd, Exeter, Devon, UK

To John

# Contents

# Figures

# Notes on contributors

**Roger Bowdler** is Director of Listing, Historic England, London, UK.

**Allan Brodie** is Senior Investigator, Historic England, Swindon, UK.

**Anya Chapman** is Programme Leader in Tourism Management, Bournemouth University, Poole, UK.

**Gary Cross** is Distinguished Professor of Modern History, Pennsylvania State University, PA, USA.

**Caroline Ford** is an Associate in the Department of History, University of Sydney, Sydney, Australia.

**Josephine Kane** is Tutor in History of Design, Royal College of Art, London, UK.

**Nick Laister** is Senior Director, RPS Planning and Development, Abingdon and former Chairman, The Dreamland Trust, Margate, UK.

**Duncan Light** is Senior Lecturer in Tourism Management, Bournemouth University, Poole, UK.

**Eleanor McGrath** is Senior Programmes Manager, Art Fund, London, UK.

**B. Nilgün Öz** is a Conservation Architect and PhD Candidate, Middle East Technical University, Ankara, Turkey.

**Bryant Simon** is Professor of History, Temple University, Philadelphia, PA, USA.

**Ian Trowell** is a PhD student, Sheffield School of Architecture, University of Sheffield, Sheffield, UK.

**John K. Walton** is a former IKERBASQUE Research Professor, Instituto Valentín de Foronda, University of the Basque Country UPV/EHU, Vitoria-Gasteiz, Spain.

**Jason Wood** is Director, Heritage Consultancy Services, Lancaster, UK.

# Preface and acknowledgements

This collection of essays brings together leading writers from a variety of disciplines to explore the social history and cultural heritage of the amusement park in the UK and in international settings. The book is timely in view of the reopening of Margate's Dreamland as the 'world's first heritage amusement park' and the proposed broadening of heritage designations in England to embrace amusement parks, including the unique collection of early rides at Blackpool Pleasure Beach. It provides a thematic, comparative exploration of the development, decline and significance of the amusement park and supplies commentary on the relationships between history, culture and heritage through a thought-provoking set of case studies which challenge imagined academic boundaries while offering important contributions to policy-making and regeneration initiatives. Rooted in the British experience but informed by extensive international coverage, the book provides an accumulation of evidence and perspectives and is intended to be a valuable resource to stimulate teachers and students of history, heritage, tourism, leisure, technology and design to exploit new sources and develop new insights into a neglected aspect of popular culture.

I would like to take this opportunity the thank all of the contributors for their enthusiasm, commitment and patience in what at times has seemed to be verging on the proverbial rollercoaster ride. I also extend my gratitude to various organisations for kindly granting permission to reproduce certain images: Blackpool Pleasure Beach; Mario Tama/Getty Images; National Fairground Archive, University of Sheffield; State Library of Victoria; State Library of South Australia; State Library of New South Wales; National Library of Australia; Koç University VEKAM Archive; Izmir Chamber of Commerce Glass Plate Negative Collection (courtesy of Fikret Yılmaz); ACFPL Atlantic City Heritage Collection; www.facebook.com/Southport-Yesteryear; Joyland Books Archive; Margate Museum (courtesy of Thanet District Council); The Dreamland Trust; and Historic England.

That the book appeared at all is due to the dedication initially of Val Rose at Ashgate Publishing and latterly of Faye Leerink and Priscilla Corbett at Routledge Research.

**Jason Wood**
*Istanbul, December 2015*

# Introduction

# 1    Fair grounds for debate

## Celebrating the heritage of amusement parks

*Jason Wood*

I was born, lived and went to school all within earshot of Blackpool Pleasure Beach. The roar of the wooden rollercoasters and screams of their passengers provided an audible backdrop to my childhood. At the time, this did not seem particularly out of the ordinary. But in a town where my walk to school took me under giant fibreglass spacemen suspended between traffic lights, nothing seemed out of the ordinary. And because children take most things in their stride, it never occurred to me to question any of this or to realise my good fortune in having a world famous amusement park on my doorstep.

A trip to the Fun House was a regular weekend ritual, perhaps followed by an exhilarating ride on the Flying Machine (at this age, I had not yet summoned up the courage to attempt the Big Dipper). But my lasting memory of these early visits is of the Laughing Man (or King as I remember him), a clownish automaton in a glass booth outside the Fun House, rotating and rocking back and forth with a mini-me Prince on his knee, both laughing hysterically (Figure 1.1). This spectacle always drew a crowd, and not just because the attraction was free. There was something compelling about the unremitting maniacal laughter that was fascinating and, after several mesmerising minutes, slightly creepy and even a little frightening. There was also an apocryphal story about the man whose job it was to turn the Laughing Man on in the morning and off at night becoming so sensitised to the laughter that on an occasion when there was a fault and the laughter stopped, he thought it was time to go home.[1]

In my mid-teens during school holidays I worked a couple of summer seasons on a concession stall not far from the Fun House and therefore within irritating hearing distance of the incessant Laughing Man, although the audio tape loop welcoming visitors to the Alice in Wonderland ride was more nauseating as it was immediately opposite the stall. This kind of seasonal job on the concessions, persuading people to part with their money, is known as a 'blagger' or 'barker' – some of my colleagues maintain that I have not stopped blagging or barking since. I had opted to work the stalls rather than the rides for two reasons. First, with the possible exception of working the dodgems, the stalls supplied greater opportunities to chat up girls. Second, you could make more money.

As the name implies, the stalls were not controlled (as they later would be) by the Pleasure Beach company so, unlike my friends operating some of the rides, I

*Figure 1.1* The Laughing Man outside the Fun House at Blackpool Pleasure Beach, 1962

Source: © Blackpool Pleasure Beach

was not a direct employee on a fixed wage. The financial benefit of my particular stall accrued in two linked ways. It was a Can-Can concession, nothing to do with a Parisian dance routine but a stall where a punter was cajoled to throw three bean bags at a pyramid of ten tin cans to knock them off a shelf and win a teddy bear. The all-important word here was 'off', which appeared in very small text on the otherwise large and invitingly simple instructions. These were designed, of course, to give the impression that a teddy bear could be yours if you only knocked the cans over. Knocking them *over* was relatively easy; knocking all ten *off* the shelf, another matter entirely. The delicacy of pointing this out to the competitive father of two or the testosterone-fuelled youth with impressionable girlfriend fell to me. (I left the stag parties and drunken Glaswegians to my much older colleague.) But handled tactfully and always with the enticement of a teddy bear in the offing, it was surprising how many more failed attempts (and therefore extra income) could be generated. Over the course of a summer's day and well into the night, this soon mounted up. Naturally, a lot depended on how much blagging and barking you were prepared to put in. And this is where an unauthorised bonus scheme came into play.

The owner, who had been in the business since the Second World War, had several stalls. He also had charts recording the takings for each stall for every day of the season for the past thirty years. He even recorded the weather. Armed with these data, the owner reckoned to know almost to the pound how much our stall should be averaging say on an overcast August bank holiday. What he had not reckoned was us obtaining a copy of the relevant charts for our stall after he mistakenly left them in the teddy bear storeroom one evening. Now that we had this information my colleague saw a money-making opportunity. Having checked the relevant day's average takings, we would work hard until this figure was reached. After that we worked even harder and the income increased exponentially. This was because, unbeknown to the owner, we were now working for ourselves and splitting the proceeds. Everyone was happy. The owner, who was not around to bother us much, was satisfied with the average takings, while we pocketed the difference.

After forty years it is perhaps a little late too confess this contribution to the black economy, and to the unofficial history of Blackpool Pleasure Beach, but having got this personal intangible heritage off my chest it is now time to set the scene for the chapters that follow.

## Development of the amusement park in Britain

The blueprint for permanent amusement parks like that at Blackpool lay in America at the end of the nineteenth century. In Britain and elsewhere the origins and development of fairground attractions can be traced back to annual fairs, pleasure gardens and especially international expositions and exhibitions, but it was the new breed of self-contained amusement parks at Coney Island in New York – Steeplechase, Luna Park and Dreamland, opened between 1897 and 1904 – that gave rise to the new concept and nomenclature (Figure 1.2). The successful formula was imitated at startling speed throughout America so that by 1906 more than 1,500 parks were in operation across the United States.

How that formula was imported and took root in Britain from the turn of the twentieth century is the theme of Josephine Kane's research. In her seminal book, *The Architecture of Pleasure: British Amusement Parks 1900–1939* (2013), based on her doctoral thesis, Kane considers the relationship between popular displays of modernity, the pursuit of pleasure for mass audiences and the spatial and architectural form of the amusement park landscape. In her complementary essay that opens this collection, Kane focuses on the period preceding the outbreak of the First World War. In less than ten years (1906–14) over thirty major amusement parks sprang up in cities and seaside resorts across the country attracting millions of visitors. For example, it is estimated that 200,000 people visited Blackpool Pleasure Beach on a typical bank holiday weekend in 1914.[2] The amusement park had quickly become a key component in the urban 'pleasurescape' of modern Britain.

*Figure 1.2* The entrance to the original Luna Park, Coney Island, 1906

Source: Illustrated Postcard Company, New York © The author's collection

As Kane explains, their appeal transcended age, gender and class boundaries, mixing fantastical architecture, technology and multi-sensory thrills in an uplifting, liberating and exciting collective experience. She argues that amusement parks became a defining counterpart to working life in the modern metropolis, promising release rather than escape from the demands of the everyday urban environment by offering a heightened, otherworldly version of it: speeding rides, repetitive mechanical noise, multi-coloured electric lights, the anonymity of transient crowds and uninhibited behaviour. This goes some way to explain why large numbers of people, predominantly drawn from the industrial and white-collar masses, were prepared to pay for pleasure rides on whirling machines that replicated, but were carefully distinguished from, their industrial and transport equivalents. She concludes that by removing the practical, physical and ethical limits of new technologies, amusement parks enfranchised the masses into an elite culture of modernity and kinaesthetic pleasures. It is especially significant that women formed a major and highly visible element of the crowd. Indeed, amusement parks may be seen as part of a wider process in which commercialised entertainment increasingly catered to the female consumer.

Nearly forty amusement parks operated across Britain in the years before the outbreak of the Second World War. The interwar period was bound up with far-reaching changes in culture, society and economy. Modernism was an architectural and broader social movement emphasising a rejection of the past and a radically reformed future. There were new architectural styles, designs and technologies to be applied to the amusement park business, new mass markets to be exploited, and new socio-political ideas and reforms to be accommodated. At Blackpool Pleasure Beach, for example, familiar rides received a comprehensive makeover in an attempt to create an overall visual unity, emphasising 'clean lines, new materials and the respectability of modernity'.[3]

Post-war, though, new more difficult challenges were encountered: the rise of inland theme parks, for example. Ian Trowell's chapter focuses on the notion of the British theme park and introduces some new concepts through instructive examination of Battersea Amusement Park and Alton Towers. This is familiar territory for Trowell, having previously worked at the National Fairground Archive where many plans, drawings, photographs and associated documents provide an opportunity to examine the construction of these parks and the bold intentions of the people behind them. His contention is that the prototype 'theme park' created at Battersea as part of the 1951 Festival of Britain in London, and what is commonly held to be Britain's first theme park at the re-branded Alton Towers in the early 1980s, should be considered as axis points in an understanding of British amusement places. He considers the spatial elements of Battersea and Alton Towers and how they functioned, in contrast, for example, with the non-spatial approaches of Kane. In relation to the first Disney venture of 1955, he explores whether Battersea was influential in fermenting Disney's ideas and if, a quarter of a century later, Alton Towers was an overt copy of Disney, or something else entirely, and indeed anti-American.

The successful introduction of the Vekoma Corkscrew rollercoaster at Alton Towers in 1980 cemented the idea of the theme park in Britain and fired the starting gun for the subsequent arms race of rollercoaster technology at other inland parks in the UK, emulating what the post-Disney generation of theme parks in America had done. But the Corkscrew was not in fact the first of the new generation of megalithic looping rollercoasters to be installed in Britain. That honour belongs to the Arrow Dynamics Revolution that opened at Blackpool Pleasure Beach in 1979, giving riders the chance to experience a single loop from a mounted start both forwards and backwards, and itself a modern steel version of the Loop the Loop (or Topsy Turvy Railway) first seen at Crystal Palace in 1902. Blackpool was the exception, however, as the increasingly extreme thrill rides struggled to take hold in UK seaside resorts. Instead, they would become intrinsically associated with those inland parks and gardens that had built upon the Alton Towers model.

## International case histories

In the first of five international case histories, the Australian cultural and environmental historian, Caroline Ford, dissects the politics of amusement in Sydney. For some years now, Ford has researched and written about the history of Sydney's beaches and of Australian beach cultures, impressively abridged on her own web-site.[4] Her chapter focuses on an amusement park that never was (Bondi Beach) and Sydney's first and only Luna Park at Milsons Point under the north pylon of the city's new Harbour Bridge, set in the broader context of the successes and challenges of other Australian Luna Parks.

The Bondi case is especially illuminating. In 1929, a proposal by the local municipal council to establish an amusement park on the famous beach divided the community. Local investors welcomed the initiative and its promises to boost the district's night-time appeal; its detractors feared it would cause visual and aural pollution and attract 'the lowest types of human vermin'. After four years of scrutiny by local media and politicians, and a formal enquiry that heard from doz-ens of witnesses, the plan came to nothing. Studying a range of archival sources, Ford explores the opponents' fears and objections with a view to understanding the ways in which ideas about public amenity, open spaces, commercialism and the aesthetics of 'water views' informed the debate and ultimately influenced pub-lic policy in the early 1930s.

Half a century later, the future of Sydney's Luna Park was itself under a cloud, the subject of a clash over lights, noise and crowds, which echoed the earlier bat-tle at Bondi. It was closed in 1979, following several safety incidents and a fatal fire. Yet, in a successful fight for its survival, it was community action followed by governmental endorsement of its heritage significance that ultimately ensured that Luna Park was retained in public ownership and conserved and adapted for continuing use as a public amusement space. So, in a neat, ironic twist, the essay comes full circle. Changing heritage values over the intervening decades now means that both Bondi Beach and Luna Park are designated heritage sites; the first recognised in effect for its heritage of resistance to an amusement park, the second for its heritage as an amusement park. Both places are driven by a strong sense of local cultural attachment and identity and, as Ford argues, understanding the his-tory and heritage of absent amusement parks is crucial to properly understanding the appeal of those which ultimately succeeded.

The chapter by the present author and B. Nilgün Öz on the early development of the Turkish amusement park is the first in English on this under-researched subject. As husband and wife, heritage consultant and conservation architect, they also feel uniquely qualified in this topic having endured the distinctly odd and frankly silly experience of being evicted from Izmir's Luna Park on their wedding day, because one of the owners objected to them having their photographs taken in front of his ride.

Much has been written on the massive project to modernise Turkey in the years following the establishment of the secular nation-state in 1923. Enjoying casual and pleasurable activities became a significant part of the newly embraced life-style, yet the amusement park rarely gets a mention and remains a neglected area of study. In this necessarily provisional case history of modernity, Wood and Öz focus on two of Turkey's earliest Luna Parks in Izmir and Ankara. Drawing on contemporary accounts, images and obscure or unpublished works, they explore the way these amusement parks came into being from the mid 1930s to the late 1950s, and the impact the urban park movement had on the proliferation of leisure facilities and amusement parks in particular.

The effects and tensions of changing ideologies under the early Republic and their reflections in modern culture and architecture find a contemporary parallel in more recent political developments such as the accession period of Turkey into the EU and particularly the rise of the pro-Islamist AK Party, under former Prime Minister and now President, Recep Tayyip Erdoğan, with its determined reversal of some of Atatürk's original socio-cultural reforms. For example, in July 2014, shortly before the presidential elections, Erdoğan's Deputy Prime Minister called for a 'ban on women laughing in public' in an overtly discriminatory speech that also deployed phrases like 'moral corruption' and 'degeneration in society'.[5] Such extreme opinion is, of course, a defining mark of autocrats and a sign of power gone – in this case hilariously – mad. Architecture, too, does not escape such aggressive ideologies. In its refurbishment, Ankara's Youth Park has succumbed to a kind of Neo-Ottomanisation with its modernist buildings now replaced or clad in Seljuk and Ottoman architectural styles. The Luna Park has so far survived this re-making, despite earlier attempts to remove it altogether, though the laughter police would be hard pressed to enforce any ban on women having fun.

The current threat to traditional Luna Parks like Izmir and Ankara is looming from what is billed as Turkey's new breed of 'international mega theme parks'. Erdoğan was on hand to open the then largest of these in 2013. Vialand has around fifty attractions spread over four themed zones, combined with a public park, performance centre, hotel and the predicable shopping mall. In more than a nod to Disneyland, Vialand welcomes visitors through a striking castle structure, leading to a 'Main Street' area with Ottoman-style shops, houses and other buildings designed to reflect daily life in old Istanbul. The Ottoman theme is continued in Fatih's Dream (*Fatih'in Rüyası*), a dark ride where passengers are invited to take 'an exciting and educational trip' through the Empire and witness Fatih the Conqueror's capture of Istanbul.[6] Vialand is now set to be eclipsed by what is claimed to be one of the world's biggest theme parks, and Europe's largest. Still under construction on the outskirts of Ankara, Ankapark will, like the Youth Park, yield to Seljuk and Ottoman architecture as the provocative style of choice.[7]

'Don't stop at Disneyland', advises a 1978 poster, 'Knott's Berry Farm [is only] minutes away and worlds apart'.[8] Knott's visitor numbers had peaked in the mid 1970s and were now on the slide, the park dropping from third to sixth in the league table of most popular amusement parks in America. But rather than making a bold marketing move against its southern Californian competitor, Knott's presents itself as 'California's second greatest attraction', and one that complements a trip to Disneyland at nearby Anaheim. Despite investment in new high-tech thrill rides, including what is claimed to be the world's first upside-down rollercoaster (the Corkscrew in 1975), publicity continued to stress the park's original tribute to the 'Old West', where 'you can ride a stagecoach, watch a wild west show, and even pan for real gold' as part of 'a nostalgic trip into the past'. This improbable amusement park in the shadow of Disneyland is the subject of Gary Cross's essay in which he explores, through meticulous use of original archive sources, the curious and contradictory history of Knott's Berry Farm and offers comparisons with other American amusement parks that emerged about the same time.

Walter Knott's roadside fruit stand and chicken restaurant of the 1930s was nothing out of the ordinary. That was until the 1940s, when it gradually acquired an Old West Ghost Town, initially as a free attraction for customers who were invited, while waiting for their restaurant table, to consume instead a curious, but obviously appealing, mix of whimsy and artefact, entertainment and patriotic heritage. Later, Knott would enclose the Farm and begin to charge admission. When Knott's youngest daughter assumed leadership in the 1970s, building two themed areas, Knott's had transformed itself into a major amusement park while retaining much of the western theming. As Cross concludes, it was Walter Knott who really led the way into the amusement/theme park era in the 1960s. Disneyland is often seen as the template but Knott's Berry Farm is more typical of the modern American park: more 'accidental', less linked to the film industry and, in contradiction to early 'uplifting' or historical intents, gradually becoming a centre of ultra-modern thrill ride technology.

The HBO period crime drama television series *Boardwalk Empire*, set in Atlantic City, New Jersey, during the Prohibition period of the 1920s and 1930s, received widespread critical acclaim, particularly for its visual style and basis on historical figures.[9] The series ran for five seasons (2010–14) and was inspired by Nelson Johnson's book *Boardwalk Empire: The Birth, High Times, and Corruption of Atlantic City* (first published in 2002) which 'traces the city's long, eventful path from birth to seaside resort to a scandal-ridden crime center and beyond'.[10] Visual effects artist Chris 'Pinkus' Wesselman worked on the series and used various photographic sources and architectural plans to research what Atlantic City looked like at this time so that he and his team of computer-generated 'imagineers' could recreate the boardwalks and seafront buildings as accurately as possible. But it was the piers that presented the biggest challenge: '[They] were one of the toughest parts because every summer they would change – new houses, new advertisements'.[11] And they were long too.

The amusements stretched from 'the New Jersey shoreline [to] somewhere near the coast of Spain' was one visitor's embellished description of Atlantic City's Steel Pier and how it jutted far into the ocean. Steel Pier, or a 1920s computer-generated version of it, made a regular appearance on *Boardwalk Empire*. First opened in 1889 it was a must-see among the attractions that made up the so-called 'Nation's Playground'. Beginning in the 1920s, it featured a myriad of different entertainments, ranging from an 'authentic' Hawaiian Village with hula-dancing 'natives' to boxing cats and the famed diving horses. It was a prized possession and one George Hamid finally got his hands on in 1945. It is Hamid, and his tenacious period of ownership through to the early 1970s, that form the basis of Bryant Simon's chapter.

Simon's book, *Boardwalk of Dreams: Atlantic City and Fate of Urban America* (2004), is an inspiring read. It may not have inspired a television series but its moving narrative tells not just a nostalgic tale of Atlantic City's rise, near death and reincarnation but reveals the boundaries of leisure and urban space in post-Civil Rights America. If Atlantic City can be viewed as a microcosm of America's cultural trajectory in the twentieth century, then Simon's essay on Steel Pier is in many ways a microcosm of the troubling fate of Atlantic City itself.

Hamid inherited the mixed-media and mixed-attraction at a time when the presence and performance of race remained central to its mass appeal. Steel Pier attracted millions of visitors every year, pandering to the contradictory impulses and desires of the vast middle of America with its wide-ranging and telling array of exhibits, minstrel shows and repeated concerts by Frank Sinatra. Using copious archival sources Simon explores Steel Pier's success by decoding what it offered, examines its decline in the 1960s, and shows how its owners attempted (and failed) to revive (through trying to recreate the racial geographies and sensibilities of the past) the amusement site in the 1970s.

The journey up Tibidabo mountain

> starts with a short train ride from the centre of Barcelona, [after which you] take the Tramvia Blau . . . half way up the mountain. If you wish to go all the way to the top of Tibidabo . . . you will also need to catch a funicular train from the half way point. Once at the top of the mountain you will be rewarded with magnificent views of Barcelona [and] an amusement park.[12]

Tibidabo amusement park is one of the oldest in the world and, despite its extraordinary mountain-top location and convoluted transport system, continues to thrive with many of its original rides still operating.

There is a reason why amusement parks tend to favour flat, easy to get to, places. While the successful longevity of Tibidabo is clearly an exception, a 'cramped, inaccessible, windswept, cloud-capped mountain site' above Bilbao, removed several degrees of latitude and longitude from Barcelona's Mediterranean climate, unsurprisingly proved too ambitious. Bilbao's answer to Tibidabo, the *Parque de Atracciones de Vizcaya* at Artxanda, opened in 1974 and closed in 1990, which means that it has now been an abandoned amusement park for almost a decade longer than it was one. In John K. Walton's chapter, this decaying vision of misplaced political dreams and lost heritage gets a thorough analysis.

The trajectory of this unique venture, 'its genesis, intended purpose, management and demise' throughout the troubled years of the transition to democracy in the Basque Country, ends with the story of its afterlife as a deserted and decaying monument to remembered or imagined pleasure and a source of inspiration for the 'commemorative creativity of artists, bloggers and online archivists'. The sense of loss and nostalgic affection for the place is palpable, particularly on websites like *Esperando al Tren* with its mystical but eerie collection of photographs of derelict rides, empty restaurants and paper-strewn offices. Like the generation of *Bilbaínos*, Walton is clearly engrossed by this 'Marie Celeste scene' and even invokes Joni Mitchell by way of a reprimand.

Artxanda joins an alarming list of modern but short-lived amusement parks left abandoned to rust and nature. Walton himself finds a parallel in the Kulturpark

Plänterwald in former East Berlin, rebranded Spreepark in 1989 after the fall of the Berlin Wall. Closed in 2001, the park has a haunted, chilling air around it. Today, it is patrolled almost as strictly as the Berlin Wall, but this does not deter groups of urban explorers regularly running the gauntlet of the private security guards, as apparently guided tours are now no longer available.[13]

Spreepark's absurdly strange tangible remains feature along with a multitude of others on numerous 'ruin porn' websites demonstrating the residual power of former amusement parks to command place attachment.[14] Many, like Artxanda, closed for financial reasons having never lived up to projected visitor numbers. For others, the end came more abruptly due to man-made or natural disasters. The world's shortest-lived amusement park is at Prypiat in Ukraine, the city that was abandoned right after the catastrophic Chernobyl nuclear disaster of 1986. The newly completed park opened for only a few hours to give the residents something to do as they waited for rescue. It has been an apocalyptic wasteland ever since. Six Flags in New Orleans, Louisiana, USA (originally opened under the name of Jazzland in 2000) was destroyed by Hurricane Katrina in 2005 and submerged under several feet of corrosive brackish floodwater for more than a month. Like Spreepark, the site is a well-known destination for urban explorers, as well as a film location. Also under water for a while, but this time the Atlantic Ocean, was the Jet Star rollercoaster at Seaside Heights on the Jersey shore of America's eastern seaboard. Its sad fate was to become one of the most iconic images of Hurricane Sandy's aftermath in 2012 (Figure 1.3). The idea of turning the stranded coaster into a tourist attraction was met with harsh criticism on social network sites.[15]

## Cultural significance, revival and heritage protection

The third part of the book focuses again on the UK, opening out debate on such questions as spirit of place, loss and change, memory and meaning, authenticity and nostalgia and regeneration and sustainability, in the context of establishing the cultural significance of amusement parks and the case for their revival and heritage protection.

A number of Britain's large coastal amusement parks closed in the 1970s and 1980s, such as Southend's Kursaal and New Brighton's Tower Pleasure Grounds, but most managed to struggle on into the 1990s. The period between 1995 and 2007, when property values escalated, saw the frequency of closures increase to several a year, encompassing some of the biggest and best known parks, with significant heritage content: Marvel's Amusement Park, Scarborough (closed 1999), Frontierland, Morecambe (closed 1999), Spanish City, Whitley Bay (closed 2000), Rotunda Amusement Park, Folkestone (closed 2005), Pleasureland, Southport (closed 2006) and Ocean Beach Amusement Park, Rhyl (closed 2007). Quite simply, the value of the land outweighed the profits the owners were generating and consequently the parks were sold to housing and retail developers. A little over twelve months later, the recession caused the property bubble to burst, so ironically many of these sites still lie vacant or semi-derelict with no immediate prospect of redevelopment because the firms that bought them went bust.

*Figure 1.3* One of the most photographed images of Hurricane Sandy's destruction was the Seaside Heights Jet Star rollercoaster, which was swept into the Atlantic Ocean after a section of Casino Pier collapsed under the force of the storm. It was demolished in 2013

Source: © Mario Tama/Getty Images

These closures were not without controversy. In Southport, for example, popular concern for Pleasureland, one of the UK's largest amusement parks, led to public protests but ultimately failed attempts at heritage designation to prevent disposal of the rare Cyclone rollercoaster built in 1937 by the celebrated American designer Charles Paige. The campaign group SAVE Britain's Heritage was 'horrified at the mindless vandalism' and demanded that 'the chainsaws must stop and English Heritage . . . be allowed on site to inspect the structure for listing', adding that 'structures such as this are of great interest in their own right, as well as holding great meaning for the many people they have given thrills and scares to'. Their call to arms and a flurry of postings on various online fora fell on deaf ears, leading to speculation that this was a case of deliberate sabotage to prevent listing and reuse by a competitor.[16] The site was largely cleared after its closure in 2006 by the then owners, the Blackpool Pleasure Beach company, and sold back to the local authority whose plans to redevelop the site as part of a broader rebranding initiative for a refined and sophisticated resort have since come to nothing. Instead, a showman has been allowed temporarily to operate a new funfair on the site whilst its future is decided.

In considering the circumstances of Pleasureland's sudden closure, Anya Chapman and Duncan Light's essay focuses on the responses from both the local community and the local authority, drawing on qualitative interview data with employees and customers and their reaction to losing the amusement park.

The chapter goes on to consider the culture and heritage of Pleasureland and, given Southport's previous ambivalent relationship with the park and the present local authority's incompatible aspirations for (re)gentrification, questions whether coastal amusement parks like Pleasureland can be developed into 'quality' tourism products, or are they consigned to become part of the British seaside's past? Without doubt, the local authority has Southport's long-term interests at heart, but its regeneration proposals, which ultimately involve erasing all traces of the town's amusement park, simply do not recognise the cultural significance of Pleasureland for many of its residents and visitors.

Another amusement park unceremoniously closed by the Blackpool Pleasure Beach company was Morecambe's Frontierland. Opened in 1909 as the West End Amusement Park, it initially featured a Figure Eight rollercoaster by the celebrated American engineer William H. Strickler. This was replaced in 1939 by the Cyclone, a 3,000-foot rollercoaster designed, like its namesake at Southport, by Charles Paige, and described as the fastest in Europe.[17] Closed in 1999, the site was largely cleared for retail development although, a decade and a half later, parts still remain vacant and derelict (Figure 1.4). Unhappily, the Cyclone was

*Figure 1.4* Frontierland's decaying Log Flume and Polo Mint Tower – a 150-foot gyro tower relocated from Blackpool Pleasure Beach, May 2007

Source: © The author

demolished, along with the iconic Noah's Ark, at the time one of only three surviving in the world.[18]

Dreamland in Margate could have gone the same way. Home to the UK's oldest surviving rollercoaster – the 1920 Scenic Railway – one of Britain's most significant and well-known amusement parks was in decline and under threat. But a proactive campaign to list the Scenic, save the site from redevelopment following closure and eventually revive it as a 'heritage' amusement park – the world's first, in fact – is an inspirational story. Undeterred by initial reluctance on the part of the local authority, obstinacy on the part of the site's owners and even an arson attack on the Scenic, the campaign went from strength to strength, buoyed and encouraged by favourable planning inquiries and repeatedly successful grant applications.

The man at the heart of this story of perseverance and eventual triumph is Nick Laister. Laister had become aware of the importance of Dreamland and its Scenic Railway due to a growing personal and professional interest in the history of British amusement parks through his work as a leading authority on the UK theme park industry. Having paid close attention to the unprecedented rate of closures of amusement parks, and consequently the increasing number of historic rides in danger of destruction, Laister's vision to save Dreamland was strikingly simple, on paper at least: acquire a representative sample of vintage rides from closed or soon-to-be-closed amusement parks, restore and re-erect them at Dreamland, and create a unique heritage attraction that could contribute significantly to Margate's regeneration and further bolster its identity as a landmark destination. Importantly, this would be more than merely a retirement home for rollercoasters; it would be an amusement park dedicated to not only preserving but to actually *operating* historic rides. But time was of the essence so, some years before the serious grant money started rolling in, Laister had to act quickly, quarrying his book of contacts and successfully acquiring several rides from Pleasureland in Southport, Ocean Beach in Rhyl and Blackpool Pleasure Beach, some of which were the last surviving examples of their type in the UK.

This was just the beginning of what proved to be a lengthy but ultimately successful battle to deliver Laister's dream. His chapter is unavoidably a personal story, told as it happened, and all the more immediate and instructive for it. It is an inspiring example of how community campaigning for redevelopment can win against short-sighted or undesired alternatives. He and his team worked incredibly hard on the Dreamland concept and delivery and should be incredibly proud of a true achievement; so it is a respectful 'Kiss-me-Quick' hats off to all those involved.

The idea of a 'heritage' amusement park raises some challenging questions within the field of heritage studies and beyond. What does heritage mean at the revived Dreamland – the tangible ride, the intangible experience of riding, the context of the ride in a landscape, or something else? Probably all three, but, of course, 'heritage' may not be the word that everyone might use to ascribe meaning. Clearly, there are tangible remains that attract the usual array of heritage terms and classifications, principally because of the listed status of the Scenic Railway and other significant assets. It is worth noting, however, that a relatively small percentage of the Scenic's original fabric survives, it having been routinely repaired and rebuilt after various fires over its 95-year existence. There is, therefore, a question mark over its authenticity in strictly fabric terms (a discussion to be picked up shortly). Perhaps its key heritage attribute, therefore, lies in its continuity as an attraction and in the authentic 'retro' appeal of the ride. Importantly, the Scenic was restored not as a museum piece but as an active ride to be experienced for real and, as no ride on the Scenic is identical (given the manual operation of the braking system), to be re-experienced. The quality of the experience and the potency of the emotional attachment to the Scenic and other historic rides are, of course, determined to a large extent by their traditional setting, so it would seem that landscape context and place attachment are also important for the overall heritage concept to work.

Does the changing context of rides moving from one landscape (for example Pleasureland, Southport) to another create any problems for people, and does it alter local identities? As Chapman and Light reveal, at Southport there were clear issues of identity and of place at play following the decision to close Pleasureland. It is unlikely, therefore, that the anger of those who felt disenfranchised by the closure would have been nullified by the knowledge that some of the rides were to be saved and relocated to Margate. People in Margate, however, are probably not as concerned about where any of the rides came from. Most of the types of historic rides that form the core of the new Dreamland did at one time or another exist at the old Dreamland. Also, in the past, rides were regularly moved from one park to another, so it could be argued that the appropriation of rides from Southport and elsewhere is a continuing tradition.

What does Dreamland mean for younger audiences who may not share in the park's cultural memory? The main target audience for Dreamland is family orientated, including the 'baby-boomer' generation with a longing to reconnect with their childhood. This familial association can be important as children often return as parents to places associated with early family holidays and cultural memories. Having said that, Dreamland will have to appeal to younger contemporary audiences, perhaps through marketing the comparison between the steel slickness and risk-averse nature of the thrill rides at today's more modern theme parks and the wooden, 'shake-rattle-and-roll' chanciness of the older types of ride. There is an element of the unknown, an element of jeopardy possibly, which might prove an attraction. This is certainly the case at Blackpool Pleasure Beach where the historic rides are enjoyed by both young and old.

Does Dreamland risk becoming a theme park to amusement park culture? The concept is to offer physical, intellectual and emotional access with an appeal to tradition, authenticity, nostalgia and identity – a living museum or landscape rather than a theme park experience. Inevitably, there will be people who will derive their memories from simulacra rather than 'real things', but only time will tell whether the 'hyper real' version of Dreamland replaces actual memory.

These pertinent questions, and others like them, were originally posed by Eleanor McGrath in a series of interviews (including one with the present author) conducted as part of her masters dissertation on Dreamland, completed five years before the park re-opened. McGrath's chapter here, based on this and subsequent research, complements that by Laister. In it she smartly deconstructs the new Dreamland heritage concept, visitor model and design vision with reference to sociological theory and a number of international comparators. The interplay of an additional interview with the creative design consultants is particularly valuable and lends her contribution immediacy.

Phase One of new Dreamland opened to the public in June 2015 with the restored Scenic Railway four months later. McGrath's intention is not to predict or anticipate the outcome of the completed project, or to scrutinise the approach of those involved to date; rather, her essay provides an overview of the debates in which Dreamland may find itself in the future, when the attraction has undergone all phases of its transformation.

Dreamland's initial driving force was that of preservation of both the tangible heritage of the historic rides and the intangible heritage of experiencing them and amusement park culture in general. An unwavering commitment to this philosophy will be important going forward, whilst also ensuring an economically viable attraction. In this regard, there is already some concern that Laister's original cohesive vision may be subject to transition and change, resulting in Dreamland occupying a potentially uncomfortable space between historical artefact and the contemporary realities of profit. The park's commercial operator has to remain conscious of the place's heritage significance and benefit, otherwise there is a danger of mission creep with the introduction of newer, cheaper and more reliable rides which require less maintenance.

David Uzzell, always worth quoting on people–environment relationships, surmises the strangeness and importance of heritage to the human psyche, and how the world is shaped within the term, when he writes:

> although the past is elusive it has a critical effect on the present and the future. It sits somewhere 'out there', a tangible resource and source of inspiration, meaning and identity as well as commercial profit. It is a physical reality that is more than just the fabrication in our minds . . . But it is clearly 'in here' as well, in the minds of the observer; it is a social construction, an empty box, waiting to be filled with our values, beliefs, desires.[19]

From an environmental psychological perspective, therefore, Dreamland is a vessel, grounded in existing memories, remains and fragments but the layers being constructed on top of it and around it will characterise the final outcome. The approach to how this is achieved will be integral to the site's future.

'There is no period so remote as the recent past' is one of the more repeatable quotes from Alan Bennett's play *The History Boys*:

> Our perspective on the past alters. Looking back, immediately in front of us is dead ground. We don't see it, and because we don't see it this means that there is no period so remote as the recent past. And one of the historian's jobs is to anticipate what our perspective of that period will be . . .[20]

What has come across strongly in recent academic discourse is the diverse range of values associated with the recent past and the complex nature of their interaction with contemporary issues. The recent past is a congested and contradictory environment; it is a 'lived in' and enlivened heritage with multiple stakeholders and multiple voices. It has yet to acquire the legitimacy of age and the consensus about its lasting value that are synonymous with more established and officially endorsed embodiments of 'heritage'. As the chapters in this part of the book have already revealed, it can be curious, strange and untidy, often intangible, linked to collective memories and entwined with the ambivalences of nostalgia. It is therefore sometimes invisible, and almost always unofficial, and largely disowned by the usual categories and criteria for protection. As a consequence, the recent past is also a contested environment, increasingly under threat from sanitising management regimes, more and more a focus for class politics and associated 'culture wars' and media inflections, and a growing area of popular concern.[21]

In addition, it often seems the historic environment sector does not know how to capitalise on public interest in heritage or to capture what matters to people and why. This requires a different way of working and a coming to terms with a different set of meanings and values of relevance to local people – meanings and values that are intrinsically linked to familial or community traditions, routines and practices, subjective and emotional attachments to ordinary places, often with potent connections to the recent past.

As McGrath acknowledges, the Heritage Lottery Fund has demonstrably helped to broaden the definition of heritage to embrace the recent past. Indeed, through its support for Dreamland and other projects in British seaside resorts there is now a growing appreciation of seaside and amusement culture, which in turn may spark demand for heritage designation.

The designation of amusement parks and fairground rides is the topic of the final chapter of this collection, in which Allan Brodie and Roger Bowdler, both from Historic England (formerly English Heritage), set out some of the challenges and opportunities ahead. Historic England is the UK government's heritage agency tasked with protecting the historic environment of England. However, a search of the National Heritage List for England (NHLE), the government's statutory designation record and a fundamental tool in protecting the country's heritage, confirms that among the hundreds of thousands of entries only a handful of amusement park rides and related structures are afforded listed building status. The first and most important ride to be listed was, of course, the Scenic Railway at Dreamland, now graded II*, ranking it among the top five and a half per cent of all listed buildings. But this is very much the exception, and Historic England realises there are other candidates worthy of potential designation. Brodie and Bowdler start by reviewing how the agency's perspective on the recent past has been evolving, principally in relation to the seaside where amusement parks are of course a key part of the story. Having examined what amusement park rides and buildings have survived in England, they then consider how these may meet the legal criteria against which Historic England must judge individual cases and, where appropriate, deliver protection through designation.

Of the determining factors that inform the designation process (for example, age, rarity, historical association, design, innovation, group value, etc.), it is perhaps the question of authenticity, an essential element in defining, assessing and monitoring cultural heritage, that is likely to be uppermost in the minds of the decision-makers. Seaside buildings are very commonly prone to change. The damp and salty environment attacks fabric, while commercial considerations, health and safety, wear and tear all lead to further alteration. Amusement park rides are on the whole high-maintenance structures, with annual programmes of repair and sometimes rebuilding. They are also especially vulnerable to fire damage. So, when does a rollercoaster, for example, with much-replaced wooden components, cease to be historic? At what point do these piecemeal alterations fatally undermine its authenticity?

Help may come from the unlikely source of Nara in Japan, home to another abandoned amusement park in the shape of the looming dystopian Disneyland that was once Nara Dreamland,[22] but home also to the *Nara Document on Authenticity* that grew out of a joint UNESCO, ICCROM and ICOMOS conference held in the city in 1994.[23] The conference and subsequent document finally broke with Western notions of heritage that perceived authenticity as a matter of form, design, materials and substance (imposed by mechanistic formulae or standardised procedures laid down in documents like the Venice Charter), and made provision for the inclusion of broader, Eastern understandings of authenticity (essentially questions of meaning, use, function, traditions, techniques, location, setting, spirit and feeling). Nara confirmed the concept of 'progressive authenticities', so that the layers of history that a cultural property acquires through time can be seen as authentic attributes of that cultural property. For instance, the Japanese conservation practice of periodic dismantling, rebuilding, repair and re-assembly of historic wooden structures can now be regarded as contributing to their heritage significance.[24] This is the context in which the authenticity of Margate's Dreamland and its Scenic Railway, and other historic amusement parks and their rides, needs to be evaluated. And it goes some way to explaining why the Scenic's listed status was upgraded after losing so much of its fabric to fire damage in 2008.

Brodie and Bowdler rightly single out Blackpool Pleasure Beach for particular attention. It is, without doubt, the pinnacle of amusement park heritage, boasting one of the finest collections of historic rides in the world. Despite inevitable losses, it is surprising how much endures at Blackpool. Worldwide, there are only thirty-five rollercoasters that pre-date 1939; the Pleasure Beach has four of them. Given that the average life span of rollercoasters seems to lie between twenty and forty years, these rare survivors stand for the far greater number of lost examples, which emphasises just how important this grouping is. Some of Blackpool's other rides have even passed their centenaries: Sir Hiram Maxim's Captive Flying Machine of 1904 and the River Caves of the following year are both relics from the time immediately before the park was formed.

As yet, none of these early rollercoasters and other rides are listed or designated in any way. Which poses two irrepressible questions: why not? And how might protection be bestowed in the future?

For many years there has been a detectable reluctance by the Pleasure Beach company to enter into a formal protected status for their historic rides or to adhere to a collecting or de-accession policy. The need to retain flexibility is at the heart of their business model and the fear has been that listing and/or legal constraints if applied may entrap what would otherwise be dynamic and changing spaces. Like Kennywood Park, Pittsburgh, USA, and Coney Island's historic Cyclone roller-coaster, Blackpool Pleasure Beach partly trades off its heritage, acknowledging that the age and significance of some of its rides deliver a unique selling point. Its self-assigned 'National Historical Marker' plaques, which provide a short history of several rides, are one level of display (Figure 1.5); the exhibit of a 1924 car from the Velvet Rollercoaster another; but clearly much more could and should be done to celebrate the special interest and careful, imaginative management of this select group.

The Pleasure Beach has nothing to fear from designation. Whether this is achieved through listing, which admittedly still carries connotations that are often misunderstood, or other levels of protection, is for others to choose.[25] For what it is worth, my own preference would be for a form of designation akin to a 'cultural landscape', one that encapsulates heritage recognition for the amusement park as a whole (as at Kennywood, for example), focusing on the sum of the parts rather than listing individual assets and acknowledging that as a living and working place the Pleasure Beach should be allowed to evolve in response to the local environment and society as it has in the past. This would permit, for instance, the relocation of rides within the park, as is sometimes required. For the time being, Historic England is actively considering mechanisms for the protection of the

*Figure 1.5*  The 'National Historical Marker' plaque for Sir Hiram Maxim's Captive Flying Machine

Source: © The author

Pleasure Beach, while at the same time accepting that it is unlikely that many more structures will be added to the NHLE. It will be a difficult balancing act and interesting to see what emerges. Whatever decision is reached it will have repercussions for other amusement parks. Hopefully, it will be reached soon. Already in the intervening period between the first listing of the Scenic Railway in 2002 and the issuing of Historic England's formal guidance on amusement parks in 2015 three major seaside parks have closed (Rotunda, Folkestone, Pleasureland, Southport, and Ocean Beach, Rhyl). At these, significant heritage assets were lost, and at those parks that remain, like Blackpool, changes may be being planned that might further erode what heritage survives.

## Towards an archaeology of amusement parks

In the *Lonely Planet Guide to Britain*'s description of Blackpool, the word 'tacky' is right up there in the first sentence, and the Australian authors soon warm to their subject with a potted history of the new industrial working class and even quote J. B. Priestley:

> Blackpool flourished thanks to the Industrial Revolution . . . In 1933, J. B. Priestley observed of Blackpool, 'To begin with, it is entitled to some respect because it has amply and triumphantly succeeded in doing what it set out to do. Nature presented it with very bracing air and a quantity of flat firm sand; and nothing else. Its citizens must have realised at once that charm and exclusiveness were not for them and their town'. There are those who visit because they have always done so, those who visit to reminisce and recapture the golden days of their youth, [and] those who visit to get pissed and get laid. [This is the *Lonely Planet Guide* now, not J. B. Priestley]. If you don't fit into one of these categories, give the place a miss. On the other hand, if you have a morbid fascination for one of the most bizarre manifestations of English culture . . . you could have a good time.[26]

As will now be apparent, I will admit to having just such 'a morbid fascination' for Blackpool and for the endearing character and cacophony of technology of its Pleasure Beach. This began with a childhood allure for the Fun House and rides like Sir Hiram Maxim's Captive Flying Machine, progressed to the odd teenage fumble in the dark in the River Caves (when the ride was not incapacitated with foam introduced surreptitiously from bottles of Fairy Liquid), to the familial rite of passage for my daughter as she was initiated to the drenching experience of my favourite ride, the Log Flume (Figure 1.6). Sadly, and for some controversially, the Log Flume is no more; removed in 2006, after forty years of operation, to make way for Infusion, a Vekoma looping coaster suspended entirely over water, relocated from Pleasureland, Southport, where it was known as Traumatizer. While the nomenclature and complicated physics of steel rollercoasters seem devised purely to extract the maximum anxiety and anticipation from the thrill-seeking, white-knuckled public, I was quite thrilled enough with the Log Flume.

*Figure 1.6* The Log Flume took existing American technology further and was the first in Britain and the longest anywhere. Here the author and his daughter descend the final slide

Source: © The author's collection

It was a poignant moment, therefore, when I stumbled across the carcass of one of the fibreglass 'logs' now high and dry in a nearby park (Figure 1.7).

If I had been paying more attention in the 'Tunnel of Love' of the River Caves, I might have appreciated sooner the ride's intention to create a virtual window into the archaeological wonders of the world. I have now been to Egypt, but not Angkor Wat or Central America, but my first sight of Tutankhamun, and so far only sight of the Cambodian temple city and architecture of the Aztec and Mayan civilisations, was from the wet seat of a Pleasure Beach gondola.

Fast forwarding to April 2010 finds me wearing my 'Dig-me-Quick' hat and organising a session on amusement park heritage at the Institute for Archaeologists annual conference in Southport, and the 'prequel' to this book. Press interest was piqued as the session coincided with a visit by English Heritage to Blackpool Pleasure Beach to assess some of its oldest rides with a view to future listing. The *Independent on Sunday* ran a story[27] while the Pleasure Beach sent a lawyer, but in the disarming shape of the then company secretary.

The idea behind the session was to challenge archaeologists to embrace the heritage of the recent past and further stimulate the overlap between archaeology and

*Figure 1.7* One of the Log Flume's simulated fibreglass logs, now recycled as a flower bed in Watson Road Park

Source: © The author

popular culture. And in this quest I started from the premise that the amusement park industry, by definition, must have an industrial archaeology. Amusement parks are no different from other expressions of industrial archaeology and are just as significant as related forms of entertainment architecture dedicated to the provision of leisure and enjoyment, such as theatres and cinemas, which on the whole receive greater recognition and statutory designation. Amusement parks, therefore, should not be immune from archaeological fieldwork and associated archival research.

By which I mean not just some chance find of a lost pearl earring dredged from the mud of the Log Flume lake below Blackpool's Big Dipper, allegedly belonging to Marlene Dietrich, where facts were not allowed to get in the way of a good story,[28] but perhaps some real archaeology where the facts might just *be* the good story. Projects, for example, at Frontierland in Morecambe or the Rotunda in Folkestone could provide an attractive way to engage with the extensive roller-coaster and amusement park enthusiast constituency and put to practical use their remarkable databases and other online resources.[29] Survey and archaeological investigation of abandoned amusement parks might well appeal to the increasingly popular community of urban explorers. Allowing access to these sites, rather than tightening security, could open up great opportunities for some innovative public archaeology, and indeed art, as was seen in the invigorating example of Artxanda.[30]

Archaeologists routinely deal with this type of challenge, if not in the context of amusement parks then on analogous sites in similar environments that present the same plethora of physical remnants, material culture, imagery and memories. It is precisely these kinds of resources that bring the potential amusement park archaeologist into partnership with the artist. In a comparable setting, archaeology and football have established a winning combination for artistic engagement. The 'Breaking Ground' project is a case point, where to some extent the archaeology of Bradford Park Avenue's decaying football ground is employed as a 'device', the ultimate aim being to filter the archaeological discoveries and interpretations through the minds of artists, using a wide range of media, and deliver a multifaceted art and archaeology project with a genuine focus on public participation.[31]

Alton Towers provided the first instance of heritage becoming art in an amusement park context. The legendary Corkscrew was replaced in 2009 but public affection for the rollercoaster led the owners to appropriate two of the helix inversions and re-erect them as a piece of sculpture at the entrance plaza. But the visual turn has now reached its postmodern, subversive apotheosis at the wonderfully anarchic 'Dismaland', a 'bemusement' park and 'sinister twist on Disneyland' organised as a group show by the anonymous satirical street artist, political activist and film director Banksy, and opened on a temporary site in the British seaside resort of Weston-super-Mare in August 2015. According to one review, Dismaland had

> all the hallmark details of a traditional Banksy event from its initial shroud of secrecy to artistic themes of apocalypse, anti-consumerism, and pointed social critiques on celebrity culture, immigration, and law enforcement ... The park is staffed by morose Dismaland employees who are uninterested in being helpful or remotely informative.[32]

\*

This has probably been an over-personalised introduction, and one my friend and former colleague John K. Walton might have admonished me for but then excused, accepting, in John's familiar adage, that 'you can take the boy out of Blackpool but you can't take Blackpool out of the boy'. John and I worked closely on the proposal for this book, which was originally intended to be jointly edited by us. We shared a lot of laughs (and Rioja), John being particularly tickled to learn there was a Disney Chair of Archaeology at Cambridge.[33] Unexpectedly, however, John retired from academic life a few months after the proposal was accepted for publication. In the circumstances it was agreed that he be released from his editorial duties and that I assume sole responsibility for bringing the book to fruition. Thankfully, John was still going to be involved as an author since, with typical promptness, his chapter on Artxanda was completed way ahead of all the others and he was happy for it to be included. As will be clear from the number of citations to John's work in the chapters that follow, he has made an immense contribution to the field of international tourism history and paved the way for the serious study of amusement parks through his authorship (with Gary Cross) of *The Playful Crowd: Pleasure Places in the Twentieth Century* (2005) and *Riding on Rainbows: Blackpool Pleasure Beach and its Place in Popular*

*Culture* (2007). Academically, it was John who strode the global stage shining like a rainbow; we lesser mortals simply hitched a ride on it. It is to John, therefore, that this book is dedicated.

True to the Walton philosophy, no 'party line' has been imposed on the contributors, and the expression of contrasting and contradictory views is to be expected. So secure all personal objects, hold on tight, and sit back and enjoy the ride . . .

## Notes

1  The Laughing Man was acquired for the 1935 season from Paris and was originally in a window of the Fun House before being re-housed in 1953. It was no laughing matter, however, when the Fun House burnt down in 1991 and the figure was destroyed, although the head survived because it had been removed for repairs: John K. Walton, *Riding on Rainbows: Blackpool Pleasure Beach and its Place in British Popular Culture* (St Albans: Skelter Publishing, 2007), pp. 62–3, 111. Since restored, but having lost his crown, the Laughing Man is no longer free or centre stage having acquired a slot machine and been relocated to a remote area of the park close to the Avalanche ride. He does, however, have an Appreciation Society: https://www. facebook.com/The-Laughing-Man-Of-Blackpool-Pleasure-Beach-Appreciation-Society-D-327509053360/ accessed 9 December 2015.
2  Gary S. Cross and John K. Walton, *The Playful Crowd: Pleasure Places in the Twentieth Century* (New York: Columbia University Press, 2005), p. 47.
3  Walton, *Riding on Rainbows*, p. 64.
4  http://carolinefordhistory.com/ accessed 11 December 2015.
5  http://www.theguardian.com/world/2014/jul/30/turkish-women-defy-deputy-pm-laughter accessed 11 December 2015. The comments prompted a social media backlash on Twitter, which the ruling party had previously tried to ban as well.
6  http://www.vialand.com.tr/ accessed 11 December 2015.
7  http://emlakkulisi.com/anka-park-martta-aciliyor/232673 accessed 11 December 2015.
8  The poster was still available to purchase on Ebay at the time of writing: http://www. ebay.co.uk/itm/351462612396?_trksid=p2060353.m1438.l2648&ssPageName=STR K%3AMEBIDX%3AIT accessed 12 December 2015.
9  http://www.hbo.com/boardwalk-empire accessed 12 December 2015.
10  http://www.goodreads.com/book/show/431030.Boardwalk_Empire accessed 12 December 2015.
11  https://en.wikipedia.org/wiki/Boardwalk_Empire accessed 12 December 2015.
12  http://www.barcelona-tourist-guide.com/en/attractions/tibidabo-barcelona.html accessed 12 December 2015.
13  Christopher Flade and Sacha Szabo, *Vom Kulturpark Berlin zum Spreepark Plänterwald: Eine VergnügungskulTOUR durch den berühmten Berliner Freizeitpark* (Berlin: Tectum, 2011); http://www.tripadvisor.com/Attraction_Review-g187323-d3515400-Reviews-Berliner_Spreepark-Berlin.html accessed 13 December 2015.
14  See, for example, http://www.atlasobscura.com/categories/abandoned-amusement-parks; http://www.therichest.com/expensive-lifestyle/entertainment/10-of-the-creepiest-abandoned-amusement-parks/?view=all; http://rare.us/story/these-10-abandoned-amusement-parks-will-give-you-chills-but-theyre-also-hauntingly-beautiful/. Siobhan Lyons is worth reading on 'ruin porn': http://www.citymetric.com/skylines/what-ruin-porn-tells-us-about-ruins-and-porn-1331 accessed 13 December 2015.
15  Ibid., for Sea Heights see http://archive.app.com/article/20121126/NJNEWS/3112 60094/Iconic-roller-coaster-Seaside-Heights-removed accessed 13 December 2015.
16  SAVE Britain's Heritage press release (20 September 2006): http://www.qlocal.co.uk/ southport/news_photo/CYCLONE_SAVE_Britain's_Heritage_is_horrified-50365229.htm accessed 13 December 2015.

17  Robert E. Preedy, *Roller Coasters: Their Amazing History* (Leeds: Robert E. Preedy, 1992), p. 36.

18  A heart-rending video about the park and its fate was published by Steven Duncuft in February 2015: https://www.youtube.com/watch?v=mBOxLCIhZNM&feature=youtu. be accessed 14 December 2015.

19  David Uzzell, 'Where is the Discipline in Heritage Studies? A View from Environmental Psychology', in Marie Louise Stig Sørensen and John Carman (eds), *Heritage Studies: Methods and Approaches* (London: Routledge, 2009), p. 326.

20  Alan Bennett, *The History Boys: A Play* (London: Faber & Faber, 2006). Incidentally, Bennett's play also helpfully provides a definition of history itself: 'History – it's just one flipping thing after another' (only he does not use the word flipping).

21  For this theme see John K. Walton and Jason Wood, 'History, Heritage and Regeneration of the Recent Past: The British Context', in Neil Silberman and Claudia Liuzza (eds), *Interpreting The Past V, Part 1. The Future of Heritage: Changing Visions, Attitudes and Contexts in the 21st Century* (Brussels: Province of East-Flanders, Flemish Heritage Institute and Ename Center for Public Archaeology and Heritage Presentation, 2007), pp. 99–110.

22  Inspired by Disneyland and opened in 1961, Nara Dreamland was permanently closed in 2006 but not demolished and is another paradise for the fraternity of urban explorers armed with cameras. For more haunting images see http://www.lovethese-pics.com/2012/03/illegal-tour-abandoned-amusement-park-nara-dreamland-65-pics/ accessed 17 December 2015.

23  http://www.icomos.org/charters/nara-e.pdf accessed 16 December 2015.

24  In ceramics, the Japanese art of *Kintsugi* (golden joinery) follows a similar philosophy, highlighting or emphasising imperfections using gold with lacquer, and treating breakage and repair as part of the history of an object rather than something to disguise. *Kintsugi* often results in something artistically 'better than new' and consequently more aesthetically pleasing and valuable: see, for example, http://www.lakesidepottery. com/Pages/kintsugi-repairing-ceramic-with-gold-and-lacquer-better-than-new.htm accessed 17 December 2015.

25  If it is decided to list a group rides, then a Heritage Partnership Agreement between the Pleasure Beach and the local authority would be a significant advantage to both parties: https://historicengland.org.uk/images-books/publications/eh-good-practice-advice-note-drawing-up-listed-building-heritage-partnership-agreement/ accessed 17 December 2015.

26  Richard Everist, Bryn Thomas and Tony Wheeler, *Britain: A Lonely Planet Travel Survival Kit* (Hawthorn, Australia: Lonely Planet Publications, 1995), pp. 693–4.

27  http://www.independent.co.uk/news/uk/this-britain/hold-on-tight-britains-amusement-parks-are-closing-fast-1941439.html accessed 17 December 2015.

28  http://www.theguardian.com/film/filmblog/2007/jan/10/marlenedietrichgirlwiththe accessed 17 December 2015. Artefacts also included a glass eye, false teeth, bra and toupee, none of which belonged to Marlene Dietrich either.

29  For enthusiast clubs in Britain, Europe and America see http://www.rccgb.co.uk/; http://www.coasterclub.org/; http://www.aceonline.org/. For databases see, for example, http://rcdb.com/; Steven J. Urbanowicz, *The Roller Coaster Lover's Companion: A Thrill Seeker's Guide to the World's Best Coasters*, rev. edn (New York: Citadel Press, 2002). For historic images see http://earlyamusementimages.com/; Martin Easdown, *Amusement Park Rides* (Oxford: Shire Publications, 2012). For the uses of archival material see for example Charles Denson, *Coney Island Lost and Found* (Berkeley: Ten Speed Press, 2002); Vanessa Toulmin, *Blackpool Pleasure Beach: More Than Just an Amusement Park* (Hathersage: Boco, 2011).

30  https://www.consonni.org/es/proyectos/luna-park accessed 18 December 2015.

31  For Breaking Ground see http://www.nevillegabie.com/wp-content/uploads/2015/08/ BREAKING-GROUND-PDF.pdf; http://www.nationalfootballmuseum.com/football-art/

residencies/breaking-ground accessed 18 December 2015. For the wider theme see Jason Wood, 'Archaeology and Sports History: Towards an Inclusive Methodology', *The International Journal of the History of Sport*, DOI: 10.1080/09523367.2015.1124862 (2016).

32  http://www.thisiscolossal.com/2015/08/dismaland/ accessed 18 December 2015. Dismaland attracted 155,000 visitors in five weeks.

33  Endowed by John Disney in 1851; no relation to Walt: John Willis Clark (ed.), *Endowments of the University of Cambridge* (Cambridge: Cambridge University Press, 1904), pp. 223–5.

# Part I

# Development of the amusement park in Britain

# 2 Mechanical pleasures

## The appeal of British amusement parks, 1900–1914

*Josephine Kane*

In April 1908, the *World's Fair* published an account of progress at the White City exhibition ground, which was nearing completion at London's Shepherd's Bush.[1] Under the creative control of famed impresario Imre Kiralfy, a series of grand pavilions and landscaped grounds were underway, complete with what would become London's first purpose-built amusement park.[2] The amusements at White City had been conceived as a light-hearted sideline for visitors to the inaugural Franco–British Exhibition, but proved just as popular as the main exhibits. The spectacular rides towered over the whole site and were reproduced in countless postcards and souvenirs. Descriptions of the 'mechanical marvels' at the amusement park dominated coverage in the national press. *The Times* reported on the long queues for a turn on the Flip Flap – a gigantic steel ride that carried passengers back and forth in a 200-foot arch – and of the endless line of cars crawling to the top of the Spiral Railway before 'roaring and rattling, round and round to the bottom' (Figure 2.1).[3] The Franco–British Exhibition was visited by 8 million people, but it was the amusement park that captured the public imagination and made a lasting impression.[4]

The following year, in a survey of London exhibitions, *The Times* acknowledged the growing importance of amusement areas, observing that: 'We do not go to exhibitions for instruction . . . the great mass of people go to them for pure amusement'.[5] The universal appeal of these amusements was deemed particularly noteworthy, and a remarkable royal endorsement in July 1909 provided definitive proof that the amusement park was not just for the masses. Queen Alexandra and Princess Victoria, visiting the Imperial International Exhibition at White City, were given a tour of the adjoining amusement park and – much to the delight of the crowds – decided to sample some of the rides. The *Daily Telegraph* reported that the Princess rode the Witching Waves (an early incarnation of the dodgems, recently imported from America), while the Queen herself took a trip on the Scenic Railway rollercoaster (Figure 2.2) and completed two winning runs on the Miniature Brooklands racetrack.[6] It was a promotional masterstroke, signalling to the country that mechanised amusement had joined the ranks of respectable modern entertainments.

But Imre Kiralfy – the brains behind White City – was far from a solitary visionary. The Edwardian era produced a number of wealthy entrepreneurs who

*Figure 2.1*  The Flip Flap and Spiral Railway at London's White City, 1908

Source: © The author's collection

recognised the huge potential for amusement parks as new forms of commercial entertainment. In 1908, the amusement park concept had been around for about a decade, but was still really a novelty. Britain's longest serving amusement park had started life on Blackpool's South Shore in 1896, inspired by the success of New York's iconic Coney Island.[7] The Pleasure Beach, as it became known, cast the die for a growing number of competitors, and the opening of London's White City coincides with the beginning of a frenzied phase of investment in American-style amusement parks in cities and seaside resorts across Britain. Between 1906 and 1914, more than thirty major parks operated around the country and, by the outbreak of the First World War, millions of people visited these sites each year.[8]

Kiralfy and his peers proclaimed themselves pioneers of modern entertainment. But did the experiences on offer really mark a significant break with the past? The early parks followed a distinct formula. Unlike their fairground cousins,

*Figure 2.2* The Scenic Railway at London's White City, 1908. Built by John Henry Iles, this ride featured scale bridges, waterfalls and mountains and was famously patronised by Queen Alexandra in 1909. Note the group of smartly dressed women waiting a turn

Source: © The author's collection

amusement parks were enclosed, fixed-site installations controlled by a single business interest. In 1903, for example, William Bean and John Outhwaite secured a £30,000 mortgage to develop 30 acres of Blackpool's shorefront into the Pleasure Beach.[9] The target audience was urban, adult and socially all-encompassing. It ranged 'from the young to the middle aged, and from those who could just afford an annual day trip, to the curious middle classes for whom the crowd itself was an essential part of the spectacle'.[10] It is estimated, for example, that 200,000 people visited Blackpool Pleasure Beach on a typical bank holiday weekend in 1914.[11]

In the interests of minimising disreputable behaviour, wardens policed the grounds and, at night, flood lighting banished opportunities for shady dealings.[12] They offered a wide range of popular entertainments, including battle re-enactments, cinema, dancing, theatres, concession stalls, landscaped gardens and often a zoo. But the amusement parks were dominated by machines for fun, and it was this aspect that marked them out as something unique. In particular, it was the rollercoaster – the defining symbol of the new parks – which enjoyed phenomenal success.[13]

Contemporary commentators were often bemused by the success of amusement parks. So, what exactly was their appeal? Why were the huge crowds – predominantly drawn from the wage-earning urban masses – prepared to pay for pleasure rides on machines that looked and sounded much like their everyday environment? The answer lies partly in the momentous cultural impact of industrialisation. The parks catered for the industrialised masses, offering – like the cinema – an otherworldly escape from the drudgery of industrial labour whilst (paradoxically) mirroring the factory system in their regularised opening times, dependence on modern transport networks and in the industrial rhythm of the attractions they offered. Just as concepts of work, time and space were altered by the onset of modernity, ideas about what constituted pleasurable experiences were transformed during the nineteenth and twentieth centuries. For people living in towns and cities across Britain, visiting an amusement park forged new understandings of modern pleasure and became a defining counterpart to life in the modern metropolis. This chapter considers the significance of the amusement park experience for Edwardian Britons, focusing on the idea of 'machines for fun' and the crowd itself to explore their enormous appeal.

## Machines for fun

The visual landscape of the Edwardian parks was quite unlike anything that had come before. Architectural eclecticism ruled. Amusement parks combined familiar styles – the exoticism and grandeur of international exhibitions and seaside piers, and the faux luxury and scenic realism of theatrical design – with the 'tober' layout of traditional fairgrounds.[14] With a single sweep of the eye, the visitor might encounter the imposing industrial skeleton of a rollercoaster, a tin-roofed hoop-la stall, the towering concrete fortress of a battle re-enactment show, a mock-Tudor house and an Indian-style tea room (Figure 2.3). At the turn of the century, eclecticism was a source of delight, a visual pleasure learned at the exhibitions and transposed to the amusement world.[15] But this seemingly ad hoc jumble was, in fact, underpinned by the visual language of machines. It was precisely this technological aesthetic – mechanical rides in motion and multicoloured electric lights – that set the amusement park experience apart.[16] At sites such as London's White City, the 'gear and girder' aesthetic of the industrialised workplace was transposed to the world of pleasure for the first time, and with great success.[17] Indeed, the bare lattice-structures and whirling mechanical apparatus of the rides played a key role in the success of the amusement park formula.

The visual delight found in machines for pleasure clearly emerges from photographic evidence of early amusement parks. One particularly arresting image, reproduced on a souvenir postcard from Kiralfy's Franco–British Exhibition of 1908, suggests the sense of pride and wonder associated with the latest thrill ride (Figure 2.1). In the foreground, smartly dressed men and women enjoy a sedate afternoon tea, their backs turned to the camera. From the formal poses and composition of the photograph, one might expect the group to be contemplating a quiet ornamental garden, or enjoying the gentle melodies of a bandstand. But, instead, the central focus of the scene is two massive and foreboding thrill machines: the aforementioned Flip Flap and Spiral Railway. A strikingly similar photograph of the Scenic Railway at Margate's Dreamland taken twelve years later suggests that by 1920 this had become a standard element of the amusement park experience. The rattle and roar of speeding carriages and screaming thrill-seekers was not, it seems, considered at odds with sedate afternoon refreshments. The spirit of

*Figure 2.3* Main Street at Blackpool Pleasure Beach, about 1923. The eclectic delights
include, from left to right: Noah's Ark (1922), Scenic Railway (1907),
Rainbow Pleasure Wheel (1912), Naval Spectatorium (1910), Big Dipper
(1923) and Helter Skelter Lighthouse (1906)

Source: © The author's collection

these images is celebratory rather than humorous or ironic, and suggests a more
complex engagement of the amusement park landscape than might at first appear.

The appeal of monumental machinery had its roots in the international exhibitions and railway and bridge-opening ceremonies of the nineteenth century, where
industrial technologies were staged as spectacle.[18] The towering rollercoasters,
swirling roundabouts and clunking revolving wheels at amusement parks visually replicated these icons of engineering progress (Figure 2.4). As the *World's
Fair* observed, 'The Great Wheel [at Earl's Court] was almost as much of a landmark for London as the Eiffel Tower is to Paris'.[19] Like the railway stations and
factories that filled Victorian cities, mechanised amusements were consumed as
rhetorical structures that demonstrated the advance of civilisation.[20] A working
drawing of the new Water Chute at Blackpool Pleasure Beach was, for example,
published in the local paper in 1907, complete with dimensions and other scientific credentials.[21]

The amusement park landscape – with the rollercoaster as its focal point – was
designed to startle and surprise, to inspire awe and wonder, to ignite people's curi-
osity, and, above all, to part them from their money. To this end, the bare lattice
structures and visible workings of cranks, pulleys and gears, had the additional
benefit of enhancing anticipation. The loading bays, often with neo-classical or
exotic facades, were designed not to beautify this machine landscape, but to entice
customers, then prime and deliver them into the realm of thrilling experience. But

*Figure 2.4* The Gigantic Wheel at London's Earl's Court, 1908. This 300-foot revolving
          wheel arrived at Earl's Court in 1896, just three years after the Ferris Wheel
          was first demonstrated at the Chicago Exposition in 1893

Source: © The author's collection

the aesthetic appeal of giant thrill-machines also lay in a combination of what
David Nye has called the mathematical and dynamic sublime.[22] Like the arrival of
high-rise buildings and transatlantic liners, these vertigo-inducing rides seemed
to defy the forces of gravity and shared the power of the railway and telegraph
to compress space and time. The landscaped or 'scenic' rollercoasters, covered
by moulded ferro-concrete mountainscapes were, for example, designed to create immersive temporal and spatial effects for the riders, rather than enhance the
aesthetic reality of the parks.[23] They created exaggerated and compressed versions
of long-distance travel and exotic locations, such as the Canadian Rockies or the
Swiss Alps. As work and travel speeded up, so an act of pleasure could be time–
space compressed into a three-minute thrill ride (Figure 2.2).

These rides created an affordable, idealised window into the world of long-distance travel. At Manchester's White City, for example, Hale's Tours – a simulator ride that featured travelogue film projected through the windows of a mock-up railway carriage – was billed as:

> more than an illusory trip, for we see the most natural pictures of all the most interesting places of resort to which the wealthy of all nations go in their hundreds and pay huge sums for the pleasure. We get it here for an infinitesimal sum of two or three coppers and the loss of only a few minutes of time, and in perfect comfort.[24]

Rollercoasters also offered, in visually accessible ways, the potential for unparalleled forms of motion: sharp turns, vertiginous inclines, even 360-degree revolution, as in the case of the Loop-the-Loop at the Crystal Palace in 1902. The verticality and sweeping curves of these rides echoed the freedom of bodily movement that defined the experience of riding them. They did not need to be beautiful in a traditional sense to be enjoyed, and the crowds were not expected to qualify them in these terms. The Franco–British Exhibition postcard illustrates perfectly how, in 1908, mechanised amusements seemed to demonstrate the advance of civilisation.

After dark, the rides and park structures were transformed by an abundance of electric lights, a celebration of the electrical sublime.[25] Illumination was rapidly embraced by amusement park owners as a way of extending hours of operation whilst, at the same time, allaying fears of criminality and sexual transgressions associated with darkness.[26] Blackpool Pleasure Beach acquired over a thousand lamps in February 1906 and, not to be outdone, Manchester's White City announced plans a month later for 'over 60,000 electric lights'.[27] An additional benefit was that the smaller rides and temporary stalls, which, by day, betrayed their cheap building materials and rapid construction, were, by night, melded seamlessly into a spectacular and entrancing display of modernity (Figure 2.5). Rem Koolhaas observed this effect in his seminal reading of the Coney Island parks. The electrified night-time landscape embodied the 'Irresistible Synthetic', an urban prototype that would later emerge in New York's Manhattan.[28]

## Speed, shocks and kinaesthetic pleasures

Visual pleasures at the amusement park were a prelude to physical engagement. In contrast to the spectacular displays at museums and exhibitions, where visitors were encouraged to look but not touch, the amusement parks were designed to be thrilling in kinetic, haptic, aural and visual ways. In 1912, Blackpool Pleasure Beach acquired the Rainbow Pleasure Wheel. A detailed description of the ride from a promotional souvenir shows how the multi-sensory nature of attractions was actively promoted. Colour, noise and speed were all incorporated in a ride which, according to its title and accompanying literature, defined the experience of modern pleasure:

*Figure 2.5* Night time view over the Boating Lake at Manchester's White City, 1910.
        White City's owner, John Calvin Brown, claimed to have installed over
        60,000 electric lights

Source: © The author's collection

> It is a Great Wheel, with two "humped" railways within the periphery, which
> is prismatically painted to represent the Rainbow. The giant circle revolves.
> The passengers are carried part of the way round, until the cars, by gravita-
> tion, run over the humps and up the other side of the Wheel; and then they
> roll back. Racing each other, backwards and forwards, through tunnels, with
> weird noises and scenes – it is Dante's Inferno![29]

Like the budding advertising industry in the early 1900s, the amusement park
landscape was designed to encourage people to pay for a thrill ride or attraction
'through processes of vision and initiation of desire'.[30] The commodification of
these multi-sensory (or kinaesthetic) pleasures played a key role in success of the
amusement park formula.[31]

The appeal of kinaesthetic pleasures was rooted in the rise of new modes of
perception in the nineteenth century. Wolfgang Schivelbusch charts the emergence of a specifically modern form of panoramic vision produced by mechanical
motion, inaugurated by the railway and sustained by the department stores and
industrial cityscapes.[32] As speed of motion causes the foreground to disappear,
the individual feels increasingly detached from their surroundings, separated
by an 'almost unreal barrier'. The landscape is thus stripped of its intensity and
is experienced impressionistically, or 'evanescently'.[33] Panoramic perception
depends on both physical speed and the commodity character of objects viewed.[34]
Schivelbusch compares the modern shopping experience with a train ride, suggesting that

the customer was kept in motion; he travelled through the department store as a train passenger travelled through the landscape. In their totality, the goods impressed him as an ensemble of objects and price tags fused into a single pointillistic overall view.[35]

Early film show how similar modes of viewing operated at the amusement park. In 1909, William Bean, owner of Blackpool Pleasure Beach, commissioned what may be the first promotional film of an amusement park, shown in Manchester to prospective visitors, and at the Pleasure Beach itself. The local paper described the film:

First a panoramic view of the whole grounds, holidaymakers everywhere, is shown. This was taken from the top of the switchback. Next comes a panoramic view of the Spanish street [with] a gay old spark, with a bevy of girls on his hands . . . Calling at the Oscillating Staircase, people are seen tumbling upstairs and down, the gay old party comes slithering down the Helter Skelter, the girls after him . . . The dash down the water chute comes out splendidly in the picture, . . . making a tremendous splash. Finally the old party slips off with his favourite girl into the River Caves.[36]

The film, produced by the New Bioscope Trading Co., included a sequence shot on board the Scenic Railway. The cameraman claimed (incorrectly) that it was the 'first film ever taken under such conditions'. The novelty of combining panoramic shots of crowds, close-up frames of rides, and filming from a moving rollercoaster was clearly impressive, and the newspaper declared it to be 'very clever' and 'a great success'.[37] The North West Film Archive holds a number of home movies from the 1920s and 1930s that attempted to capture the park landscape from within a moving rollercoaster, suggesting that new perceptual experiences formed an important and lasting component of the amusement park pleasure formula. The visitor experienced the park as an ensemble landscape of commodified pleasures, infused with speed: multi-directional crowd flows, the movement of ride machinery, and the body itself in motion.

While the *visual* experience provided by a speeding ride might be similar to a train journey, the bumps, jolts and twists of a rollercoaster offered a very different *physical* experience. How was it that being rushed up and down terrifying inclines, spun into a dizzying haze and turned topsy-turvy came to be seen as enjoyable? The answer lies partly in the cultural impact of urban modernity. By the turn of the twentieth century, the speed of travel and urban life had become normalised. The well-documented anxieties and bewilderment expressed by early train passengers and city dwellers in the mid-nineteenth century receded.[38] Travellers and urbanites became desensitised to their environments, acquiring what has been called 'the industrial consciousness'.[39]

The idea that people develop a protective mental layer against the over-stimulation of modern life was first formulated by German sociologist Georg Simmel, who observed the 'blasé' attitude of urbanites in his seminal essay 'The Metropolis and Mental Life' (1903)[40] and, later, by Sigmund Freud's 'stimulus-shield' theory in *Beyond the Pleasure Principle* (1922).[41] Simmel and Freud suggested that only extreme shocks could penetrate this protective psychological layer. Just as the amusement parks were becoming more popular, the potential of shock to be pleasurable was gaining recognition.

Writing about the Berlin Trade Exhibition in 1896 (the same year that William Bean registered a London syndicate to operate the rides on Blackpool's South Shore), Simmel argues that the modern urban experience 'produced a thirst for yet more amusement'.[42] The blasé attitude, characterised by 'an incapacity to react to new stimulations with the required amount of energy', paradoxically led the urbanite to seek out ever-new attractions.[43] Simmel observes 'the craving today for excitement, for extreme impressions, for the greatest speed of change . . . the modern preference for "stimulation" as such in impressions, relationships and information'.[44] The indifference and isolation induced by living in modern cities caused an inner restlessness that people sought to satisfy through intensified experience: 'the lack of something definite at the centre of the soul impels us to search for momentary satisfaction in ever-new stimulations, sensations and external activities'.[45]

The amusement parks, with their mechanised thrill rides and spectacular displays, were understood as an antidote to desensitisation, and so their emergence was perceived by contemporaries (both advocates and critics) as inextricably linked to the condition of modernity.[46] In 1912, the *World's Fair* reported that:

> Blackpool hungers and thirsts for novelty . . . When the Lancashire operative goes to Blackpool . . . he puts behind him the monotony of routine and yearns for novelty, sensation and excitement. The immense popularity of Blackpool's big pleasure beach provides striking proof of this.[47]

The *World's Fair* reiterated the belief that amusement park success depended on satisfying the modern person's insatiable appetite for novelty: 'only the weirdest sensations are favoured by the public to-day'.[48] Rides were thus stripped of all sensory buffers in order to re-inject the sense of velocity and danger that had been dampened by upholstered, enclosed railway carriages.[49] Olympia's Canadian Toboggan, for example, promised to 'bump with as much violence as if you were in a motor car on a bad road'.[50] The opportunity for interaction with strangers and physical intimacy on rides such as The Tickler or in the quiet darkness of the River Caves compensated for the indifference of the city street. The amusement parks represented a unique space in which the rules of social convention – reserve, indifference, class distinction – could be flouted and the stimulus-shield of modern life might be momentarily cast aside.

The search for intense experience – thrill-seeking – was understood as a defining characteristic of the modern psyche. In 1908, a journalist provides a glimpse of Edwardian attitudes to the thrilling pleasures offered by the amusement park. Thrill is described as an 'ecstasy of excitement' which 'stirs his blood, excites his brain', offering transcendent possibilities. On the Scenic Railway, we are told, even the 'mildest of men' becomes a 'reckless hero' and 'staid old ladies . . . frisky maidens'. The perception of danger and speed is essential for this momentary catharsis, enabling the individual to take 'the brake off himself' or to 'relieve her feelings'. Thrill-seeking itself was, of course, not new in 1900. But mechanically produced amusement park thrills were understood as a scientific phenomena in an era of progress. The rollercoaster ride is 'a psychological revelation' in which 'the modern man . . . enjoys primitive emotions in a scientific fashion'.[51]

The perception of thrill as an enjoyable experience depended entirely on the trust placed in the safety of the rides themselves. Despite sharing the same technological vocabulary, amusement park machines were carefully distinguished from their industrial and transport counterparts.[52] While serious and sometimes fatal mishaps frequently *did* occur at amusement parks, most were caused by passengers misusing rides – standing up in cars or leaning out. At the inquest into the death of 19-year-old Alfred Butts on the Figure Eight rollercoaster at Cleethorpes, for example, the coroner passed a verdict of accidental death following witness accounts of Butts' behaviour: 'When they neared the bottom, Butts rose a little and put his hands in his pockets, leaning back while he did so. He lost his balance then, and went over the side of the car, trailing along for a little way'.[53]

The caution demanded in daily life on the construction site, the factory floor or a traffic-filled street was evidently not translated to the amusement park – partly because the otherworldly landscape discouraged such comparisons, and partly because the concept of 'health and safety' was still very much in its infancy.[54] Machines for pleasure were perceived as safe – providing shocks without trauma – and this became a mark of progress itself. Even accidents caused by machinery failure appear to have caused relatively little concern and, in some cases, actually added to the success of a ride. Take, for example, the first serious accident on Blackpool Pleasure Beach's Scenic Railway at the height of the 1911 summer season. A car loaded with 25 people was 'thrown violently off the tracks', causing six passengers to be severely injured. The aftermath of the incident caused great interest amongst the crowd, becoming a ghoulish spectacle in its own right. A report stated that 'assistance to the injured . . . was greatly hampered, and the efforts of the ambulance workers and others hindered, by the crowd of people who immediately collected around'. Just two hours later, 'the service of the cars was resumed and they were as freely patronised as ever'.[55]

The amusement parks became crucial loci for the commodification of risk, both through the entertainment value of apparently safe thrill rides and the high-risk antics of daredevil stunt performers, and the hazards faced by park workers who operated the rides. Reports in the *World's Fair* of horrific injuries and fatalities suffered by ride operatives is testament to a level of peril unseen by visitors.[56] As Arwen Mohun observes, visitors to the amusement parks paid to avoid risk, or to watch skilled entertainers taking it.[57]

## Wonders of the modern age

For critics of modern amusement, the appeal of amusement technologies that mirrored working life was unfathomable. A writer for the *London Standard* called

Blackpool a 'pleasure factory by the sea'. Observing the 'exorbitantly crushing demands made upon [Lancashire workers'] endurance by the heavily capitalised organisation of pleasure', the writer comments that 'one may wish that the opera-tives of Lancashire would prefer . . . rustic pleasuring, though in view of their lives the year round, the wish is hopeless'.[58] So what made visitors to amusement parks, predominantly drawn from the industrial and white-collar masses, prepared to pay for pleasure rides on machines that replicated their working lives? Many elite commentators failed to grasp the clear distinctions visitors made between what they might 'endure on a day-to-day basis and what they could selectively pay for'.[59] More importantly, the amusement park with its machines for fun offered the working masses unprecedented opportunities to participate in a shared culture of modernity.

The rollercoaster, in particular, seemed to epitomise modern pleasure. This was not because the technology itself was new (early rollercoasters were essentially a variation of well-established railway traction systems and bridge construction) – but because it signalled the arrival of technology for fun. A 1906 article in the *Manchester City News*, reporting on amusement parks in Canada, described 'a bewildering maze of switchbacks, aerial flights, water chutes, scenic and toy railways' as 'triumphs of modern civilization . . . all brilliantly lighted by elec-tricity'.[60] A souvenir brochure from 1912 declared Blackpool Pleasure Beach 'a revelation of the Age of Science'.[61] The industrialisation of amusement seemed to represent how far civilised (Western) societies had progressed – modernity had reached all aspects of life, including the notion of pleasure itself. Just as shopping in a new department store, using a bicycle, or visiting a cinema were identified as activities unique to the modern age, riding a rollercoaster became one way in which contemporaries might achieve the status of 'being modern'.

Moreover, the ups and downs, sudden twists and the exhilaration of a roller-coaster ride soon became a familiar metaphor for the disjunctive and transient nature of life in the modern city. Sequences shot by mounting a camera on mov-ing rollercoasters (like that used in the 1907 film of Blackpool Pleasure Beach) were later used in commercial films as an allegorical device to denote the modern condition. In 1927, Walter Ruttmann interwove first-person shots from a roller-coaster into the narrative of *Berlin, Symphony of a Great City* (1927) to suggest the 'dizzying, frenetic vortex' of modern metropolitan life.[62] In the same year, the British film *Hindle Wakes* employed a lengthy sequence on the Big Dipper at Blackpool Pleasure Beach as a narrative turning point, sparking a scandalous 'modern' love affair between a factory girl, Fanny Hawthorne, and the factory-owner's son.[63]

From the beginning, the parks contained powerful representations of the new-est era-defining technologies, including the aeroplane, the submarine and the motor car. Sir Hiram Maxim's Captive Flying Machine, first exhibited at Earl's Court in 1903 and then at Blackpool Pleasure Beach the following year, provided a simulated taste of what it might feel like to pilot an aeroplane (Figure 2.6).[64] Having made his name as an engineer of machine guns, Sir Hiram devised the Flying Machine as a fund-raising initiative to support his ongoing experiments in

*Figure 2.6* Sir Hiram Maxim's Captive Flying Machine at Blackpool Pleasure Beach, about 1904. One of the earliest thrill rides to be constructed at the Pleasure Beach, the Flying Machine still operates on the same site today

Source: © The author's collection

aviation.[65] As the ride revolves on a 30-metre steel pole, ten suspended carriages fan outwards, creating the illusion of flight. Its arrival at South Shore marked a turning point in the development of the Pleasure Beach: it was 'one of the first indications of a new era in mechanical contrivances' which helped transform the ad hoc entertainments into a fully fledged amusement park.[66]

This novelty ride was a remarkable symbol of technological progress. The realisation of powered flight (achieved by the Wright Brothers just a few months after the Flying Machine opened at Earl's Court) was viewed as the epitome of modernity, a herald of unimaginable change.[67] The immense popularity of Maxim's Flying Machines around the country spawned various imitators. In 1909, Manchester's White City promoted its Aeroflyte as a flight simulator open to all:

> Every man would like to enjoy even for a moment the supposed sensational trip through the air . . . It is not possible for many in these days to obtain this opportunity, but the next best thing that is offered them is a short flight on some contrivance or mechanism that will produce similar sensations . . . The Aeroflyte . . . gives to the occupant of the chair the exact same sensations that are experienced by the balloonist or aeroplanist.[68]

The transcendent possibilities of flight was not the only culture-changing technology to be celebrated at the amusement park. In June 1907, Blackpool Pleasure Beach acquired an attraction that simulated a submarine descent, complete with 'scientific lecture'.[69] Later that year, Charles Cochran's Fun City at Olympia heavily promoted a similar ride – Voyage on a Submarine – which fused science-fiction fantasy with technological utopianism, playing on the transformative potential of this newest form of transportation. It was, in reality

> a sort of '20,000 leagues under the sea' illusion. You get in, the hatches are screwed down, and then the boat seems to be going down, down, down until you find yourself at the bottom of the sea among the coral and the mermaids.[70]

In 1870, Jules Verne's novel *20,000 Leagues under the Sea* popularised the idea of submarine transportation, igniting the popular imagination and fuelling the activities of engineers worldwide. The turn of the twentieth century marks a pivotal time in the development of submarines, with the French and United States navies leading the way. The race to develop submersible technology was viewed with a sense of national urgency and it is no coincidence that simulation rides appeared in amusement parks in the following decade.[71] Underwater travel was viewed as a significant break with the past, as a sign of the progress of the civilised world. The submarine was successfully appropriated by the amusement parks not just because it symbolised technological modernity and Britain's continued naval and imperial prowess, but also because – like the airplane – it offered an experience unimaginable to earlier generations.

In 1907, motoring was a new mode of transport favoured by the fashionable elite. But the opening of the world's first purpose-built racing track at Brooklands in Surrey in June of that year created a surge of popular interest in motor racing as a novel (albeit socially exclusive) sport. Brooklands signalled Britain's arrival as a racing nation, and established driving itself as an aspirational metaphor for the modern age.[72] Within a year of its completion, Blackpool Pleasure Beach had opened its own version: 'a motor-racing track that provides the delights and the thrills of a miniature Brooklands, with none of its dangers'.[73] The ride consisted of three cars, each seating four passengers, which raced along half a mile of parallel tracks at speeds of up to 12 mph, controlled by the driver.[74] The accessibility and safety of the ride were touted as particularly appealing features. A local paper reported that 'ladies can drive these cars just as well as the sterner sex', whilst 'accidents of any sort are quite out of the question'.[75]

Amusement parks around Britain quickly followed with their own versions of the Miniature Brooklands, including the ride famously endorsed by Queen Alexandra at London's White City in 1909. Patent after patent of mechanical riding devices inspired by the motor car were announced in the *World's Fair*. One of the earliest was Mr Fred Harrison's rollercoaster, Looping the Loop in a Motor Car, unveiled in 1906.[76] The Rolling Motor Track of 1910, in which three cars steered themselves around an oscillating track, is another example of the numerous designs exploiting the allure and novelty of motoring.[77] Driving in 'real life' was deemed highly unsuitable for women and beyond the financial reach of most men.[78] Rides such as these capitalised on the novelty and socially aspirational appeal of the motor car whilst simultaneously removing the physical, moral and economic constraints.

So, the amusement park appropriated cutting-edge technologies which, in the eyes of contemporaries, marked a clear break with the past and underpinned 'the modernist storyline' of the onward march of progress.[79] In the first decade of the twentieth century, these technologies – the airplane, the submarine, the motor car – were highly potent emblems of modernity, which lay beyond the reach of all but a select few. By removing the practical, physical and ethical limits of new technologies, the amusement parks enfranchised the masses – and, astonishingly, women – into an elite culture of technological modernity.

The amusement parks employed the language of 'wonder' to describe new attractions with striking regularity.[80] Blackpool Pleasure Beach's Sea Circus (an aquatic roundabout) was portrayed, for example, as 'an elaborate piece of mechanism, having many hidden wonders'.[81] Likewise, the submarine ride was an opportunity to experience 'the countless wonders of the submarine world'.[82] By describing new rides as modern wonders, the amusement parks tapped into a general fascination with (and fear about) technology itself and the dramatic changes it heralded.[83] But the assimilation of new inventions into the recreational experience of the general public also served to demystify them. Just as the national press eulogised Britain's engineering and technological leadership, providing a sense of 'collective purpose' for innovation, rides like the Flying Machine and Miniature Brooklands helped dampen ambivalence to new technologies and create a national culture conducive to technological advance.[84] At the same time, the amusement parks themselves became part of the landscape of modernity.

## The crowd

The throngs of people who patronised the amusement parks were as much a part of their appeal as the over-sized mechanical attractions. The tea gardens at Manchester's White City were carefully positioned so that its patrons were able to survey 'the constantly moving and changing human panorama as it passes along the promenade'.[85] Commentators remarked on the novelty and spectacle presented by such gatherings of people. For one journalist writing in 1907, Blackpool's amusement park attractions were overshadowed by the sheer volume of people at the Pleasure Beach. 'On the fairground', he reported, 'the spectacle was simply bewildering. One gazed in amazement, and wondered where all the people came from'.[86]

The scale of the crowds was partly a consequence of the broad social appeal of the amusements. Rides and shows at the Pleasure Beach generally charged between one and three pence, making them within the reach of all but the poorest sections of society.[87] Even London's White City – where rides charged between six pence and a shilling – a large proportion of the Bank Holiday crowd in 1908 was made up of a spectrum of industrial workers. *The Times* reported that:

> The Cooperative Societies of Newcastle, Manchester, Liverpool, Derby, Lincoln, Retford, and Hucknell each sent large parties, and, in addition, there were parties of engineers from Newcastle and Bristol, gasworkers from Cardiff, steelworkers from Sheffield, foundry-workers from Birmingham, and railway *employés* from several centres.[88]

Nevertheless, the amusement park crowd was considerably more diverse than other commercial entertainments aimed at the masses. To a far greater extent than the music hall and public house, the parks attracted equal measures of women and children. A colourful description from 1907 describes the eclectic mix who patronised the Blackpool Pleasure Beach sideshows and who ranged from 'the bewildered miner' to 'smirking young ladies, awkward hobbledehoys, self-conscious matrons, reluctant papas, and uneasy family groups'.[89] Indeed, the appeal of mechanised amusement transcended divisions of age and gender, as well as class – although debates concerning the role of mechanical amusement in the later twentieth century have certainly obscured this fact.[90] The Edwardian amusement parks – with the help of much-publicised royal and government endorsements – were consumed and enjoyed across the social spectrum.

Moreover, from the start, park entrepreneurs, keen to reproduce the success of exhibition amusements, aimed to attract a prosperous and educated audience, and clearly targeted the middle classes in their promotional material. Various strategies were employed to this end. First, comparisons with London exhibition sites were repeatedly made, with the implication that the amusement parks offered superior and respectable attractions suitable for a more refined audience. In 1907, Southport's proposed amusement park would 'combine the best features of Crystal Palace and Earl's Court'.[91] A visitor to Blackpool Pleasure Beach reported that 'we all rubbed our eyes, and asked each other were we dreaming, or had we, by some mysterious means, been suddenly transported to Earl's Court'.[92] In 1911, the newly renamed Luna Park advertised itself as the 'White City of Southend'.[93]

Second, the educational and artistic merit of attractions was heavily promoted. Entertainments celebrated historic events – in the case of Blackpool Pleasure Beach's Monitor and Merrimac Naval Spectatorium, the first battle between two 'ironclad' warships off the coast of Virginia in 1861.[94] These spectacles claimed to be authentic in every detail, and often incorporated some form of educational commentary. The submarine ride, for instance, was accompanied by a 'capital scientific lecture on the diver, his equipment, and work in the depths of the sea'.[95] Artistic merit was equally stressed. The publicity for an illusion show called 'Sculpture Bewitched' informed potential visitors to the Pleasure Beach that it was the creation of 'Mr. Hudson, a portrait painter, whose work is of such merit as to have secured his admission to the Royal Academy'. The show was 'a genuine novelty, of great refinement'.[96]

Finally, the discourse of health was employed in order to distinguish the amusement parks from other working-class entertainments. Accordingly, in 1907, Blackpool Pleasure Beach emphasised its 'clean and honest amusements'.[97] The following year, it was described to Manchester readers as 'a vast outdoor entertainment resort which skirts the sea shore [and] is completely exposed to the healthful breezes that sweep from the west'.[98] Other parks were more explicit in laying claim to the morally improving aspect of the healthy entertainments on offer – a useful strategy for quashing local opposition to new ventures. The amusement park proposed at Shoreham in 1907 would consist of 'a great variety of the very healthiest entertainments', aimed at giving 'our toilers the opportunity to enjoy a 'real bank holiday' away from the beer house and gin palace'.[99] Edinburgh's Marine Gardens, which opened in 1910, was described as a place 'of innocent amusement', which provided 'counter attractions to the public-house'.[100] By invoking the tenets of fresh air and respectable pleasures, the amusement parks appropriated the language of the rational recreation movement of the nineteenth century.[101] In addition to calming fears of bawdy and morally degenerative behaviour, they hoped to draw in women, children and wealthier holidaymakers.

The effectiveness of these strategies in attracting a broad spectrum of visitors may be gleaned from accident reports (which stated age, gender and occupation of injured parties), the contemporary press and photographic evidence. While much of the amusement park crowds were made up of the wage-earning masses – which was in itself a highly stratified group ranging from factory employees to white-collar and skilled workers – it is clear that the amusement parks were not exclusively male, adult or working class.

Amusement parks heavily promoted their universal appeal, irrespective of age. Blackpool Pleasure Beach's advertisement in 1907 declaring 'A New World. Everything Good for Young and Old' was typical of the claims made by other parks.[102] The success of such promotional rhetoric is borne out by archive evidence. In 1906, the manager of the Pleasure Beach's Aerial Flight testified in a personal injury claim heard at the Blackpool County Court that 'people of both sexes up to sixty years of age went on it without accident'.[103] In 1911, a party of elderly ladies were reported enjoying the delights of the amusement park with 'youthful enthusiasm'. 'Two giddy old dames of over 70 years of age' were whirled off the Joy Wheel, whilst another 85-year-old 'derived the keenest enjoyment from the thrilling rush round the Velvet Coaster'.[104] The amusement parks were aimed primarily at the spending abilities of a mixed adult audience, but children and families were an important constituent in the amusement park audience.

In contrast to the male-dominated venues that had previously characterised popular entertainments, women formed a major and visible element of the crowd. Far from taking a backseat, preferring the quieter gardens or more sedate attractions (and contrary to the expectations of the time), female visitors of all ages were as likely to head for the large-scale thrill rides as men. In 1910, for example, *The Times* reported a Lord Mayor's Court action to recover damages for injuries sustained on White City's Spiral Railway. The Plaintiff, Mrs Blanche Dunn, was the wife of a veterinary surgeon from Poplar, London. There is no hint in the report that Mrs Dunne, as a respectable middle-class woman patronising a mechanical thrill ride, was considered exceptional.[105] Indeed, by 1912, the manager of White City could confidently state that 'women far exceed men in the numbers patronizing the newer sensations' such as the Screamer, Flip Flap and Mountain Railway. 'Their attitude to these novelties suggests that women are certainly more enterprising than men in collecting new sensations'.[106] Given the highly restricted nature of commercial recreations available to 'respectable' women in the Victorian and Edwardian period, it is hardly surprising to find that women made up a significant portion of the amusement park's clientele. Indeed, the amusement parks, like the

cinema, may be seen as part of a wider process in which commercialised entertainments increasingly catered for the female consumer.

The amusement park crowd must be distinguished, however, from the everyday
hordes of the modern urban street, identified by Simmel.[107] The daily encounter
with the modern metropolis caused, according to Simmel, a unique psychological
adaptation in the city-dweller.[108] In order to cope with the ceaseless barrage of
sensory stimuli, urbanites attempted mentally and emotionally to distance themselves from their environment. One of two responses resulted from this attempt:
agoraphobia and hypersensitivity in extreme cases or, more commonly, indifference towards human relations – the blasé attitude.[109]

By contrast, the amusement parks promised release from the demands of everyday life, and played host to a mass of individuals joined together in the pursuit
of fun. To be part of such a collective could, as one writer described, be uplifting,
liberating and exciting – a far cry from the indifference and distrust displayed by
Simmel's urban crowd:

> You wander in search of adventure, and you find it in canvas booths, in the
> shower of sand, in the rumble of wheels, in the glad cry of the triumphant
> tripper, in the shrieks of maidens, in the glorious crescendo of a summer
> crowd climbing to the knowledge of holiday happiness.[110]

For some historians, the concept of the carnivalesque helps explain the behaviour
of the crowds drawn to the amusement park.[111] And, yet, there is strikingly little
evidence of the wild and hedonistic behaviour associated with the Bakhtinian
crowd.[112] In 1907, for example, only two cases of drunkenness were reported
at Manchester's White City during a season in which over 750,000 people visited.[113] In 1913, London's White City claimed 'there had never been a single case
of disorder of any kind' in the five years since its opening.[114] Unruly behaviour
undoubtedly manifested itself at popular resorts such as Margate, Southend and
Blackpool, but it is much easier to locate in the liminal spaces of the beach, pubs
and ad hoc seafront entertainments than in the carefully regulated amusement
parks.[115] A letter published in *John Bull* in 1909, for example, expressed outrage
at the behaviour of 'hobbledehoys and wenches . . . in the lanes, in the shelters, on
the sandhills' of Blackpool, but made no mention of the Pleasure Beach.[116]

The amusement park landscape, with its myriad of attractions, created an
atmosphere of collective freedom in which the formality of official, working life
was relaxed. But, far from representing 'a second life'[117] the crowds' experience
was framed by familiar rhythms of sociability, celebration and consumption.
Rather than turning the 'world inside out', as Bakhtin would have it, the amusement parks magnified the positive and festive features of everyday life.[118] Thanks
to new mechanical forms of pleasure, crowds enjoyed the 'holiday mood' rather
than the carnival spirit. The freedom of bodily movement, social mixing and compulsory screaming that occurred at the amusement park – but would have been
quite unacceptable in everyday life – might be seen as elements of carnival had
they not been regulated by the rhythm and movement of the mechanical rides and,

to a great extent, by the crowds themselves. Bakhtin states that carnival 'is not a spectacle seen by the people; they live in it, and everyone participates'.[119] And yet, at the amusement park, spectatorship was a key element of the amusements on offer. Members of the crowd were encouraged to be both actors in, and spectators of, the entertainment.

Archive photographs show how crowds gathered to watch mechanical rides in operation. Blackpool Pleasure Beach's Joy Wheel – a spinning circular platform on which people sat to be thrown outwards by centrifugal force – was designed with a large raised circular gallery on which people could stand to watch and laugh at the fate of those being spun around.[120] The idea was to engineer a total loss of bodily control amongst the riders – men and women of all ages – for the entertainment of spectators. One journalist described the effect:

> You may go feet first, head first, or sideways like a crab. You may go on your elbows, your ankles, the knuckles of your hands, the broad of your back, the pit of your stomach; you may go even on your eyebrows or on one ear . . . The world is full of flying arms and legs and spinning bodies until the Joy Wheel is spinning empty and triumphant [and] the arena is rocking with laughter.[121]

Rides such as the Joy Wheel show how a visit to the amusement park involved its own set of coded behaviours, ritual practices that lay somewhere between the everyday and the liminality of the beach or fairground. The lack of carnival spirit should not be taken as evidence of the suppression of popular practices of resistance which – as John F. Kasson has argued in reference to Coney Island – created 'passive acceptance of the cycle of production and consumption'.[122] The meaning of the amusement park experience for visitors themselves was rather more complex.

## Utopias at the amusement park

For people living in towns and cities across Britain, visiting an amusement park became a defining counterpart to life in the modern metropolis. Rather than an escape from the urban spectacle, the amusement park offered a heightened version of it: speeding rides, mechanical noise, electric lights and the anonymity of flowing crowds. Darren Webb's revealing analysis of Blackpool Tower suggests how framing the amusement park landscape as a utopian text might help explain their appeal to a metropolitan audience in the early twentieth century.[123] Enclosed and clearly separated from the outside world, the amusement park engineered an immediate sense of otherness, heightened by ornate entrances and clearly marked boundary lines, and by the fantastical designs of shows inside (Figure 2.7). Attractions such as the Scenic Railway, Hale's Tours, the River Caves, and various 'native' villages, emulated foreign landscapes and provided (like the Tower interiors) 'a succession of glimpses into the exoticism of other extant realities'.[124]

Defined by fantasy on the one hand, the amusement park simultaneously celebrated the very real emancipatory potential of the present. In particular, the

*Figure 2.7* Inside Manchester's White City, 1910

Source: © The author's collection

replication of cutting-edge technologies in popular rides – the flying machines, motor-racing tracks and submarine rides – testified to the 'possibilities of the future' and the ongoing advances of science.[125] Alongside these realisations of the exotic present and idealised future were nostalgic representations of the past: Ye Olde Englishe Street at Blackpool Pleasure Beach, Old London at London's White City, a medieval ruin at the Hall-by-the-Sea in Margate. These temporal elements collided at the amusement parks but in a way that encouraged visitors to 'decode the interrelatedness of their immediate present'.[126]

The sense of utopian otherness is graphically illustrated by a description of a day at Blackpool Pleasure Beach in 1910. The amusement park is presented as a place 'where life moves so swiftly and noisily, where fatigue is an unknown word, and where joy is served in deep draughts that knows no satiety'. The intensity of experience offered at the amusement park warps the space and time of every-day life: to spend an afternoon there is 'to have lived many years between noon and sunset'. It provides transformative encounters that are both revelatory and rejuvenating: 'before I went on the Joy Wheel, I had not lived. I had not drawn back the veil which secretes true happiness. To go on the Joy Wheel is to be born again; born in gaiety and baptised in the waters of irresponsibility'. The past (and the geographically distant) is presented up close in the Monitor and Merrimac, a 'theatre-like palace' where you 'learn how the Monitor and Merrimac fought their great battle off the coast of Virginia' and 'feel that you are looking across a mile of water watching naval history in the making'. Ultimately, the Pleasure Beach is a place where 'nothing is impossible' and 'freedom and forgetfulness' reign.[127]

The amusement park landscape, with its combination of fast-flowing crowds and spectacular rides, represented the pulse of a romantic vision of modern life: visceral, intense and stripped of the banality of everyday industrial labour. In doing so, these sites strove to create a kind of commodified utopia with potentially universal appeal.

## Conclusion

Amusement parks flourished not because they were vehicles of indoctrination or sites of resistance for the masses, but because they were the source of a new kind of pleasurable experience that captured a pervading sense of living through an era defined by permanent and man-made change. White Cities, Pleasure Beaches and Luna Parks offered a heightened version of the urban spectacle: speeding rides, repetitive mechanical noise, multi-coloured electric lights, transient crowds and uninhibited behaviour. Rather than offering a space of escape, the particular form of mechanical multi-sensory pleasure consumed at the early parks became a defining counterpart to city life and played a key role in making sense of the experiences of popular modernity. While critics berated the similarities between the industrial workplace and the mechanised amusement parks, for the patrons themselves the experience was far from routinised or pacifying. A visit to such a place was a treat, somewhere to go once or twice a year. Moreover, as accident reports reveal, pleasure-seekers were continually experimenting with their own ways of bringing novelty and excitement to the rides.

The amusement park offered a redefined notion of pleasure in which *doing* was as important as *watching*. Rides and attractions transformed the visitor into racing drivers, pilots, explorers, comedians, even stars of the screen. This was a form of pleasure defined by participation and, in this way, the parks provided a momentary escape from the anonymity and indifference of urban life characterised by Georg Simmel. The amusement park catered for a shared desire for sensuous and immediate engagement with life, a desire seen as a key point of tension in the mechanised age.

## Notes

1  *The World's Fair* is a national amusement trade newspaper, published weekly from June 1904. Providing news and commentary about the industry, it is read by fairground and amusement park operators across England, who use its pages to buy or sell rides and equipment, to advertise jobs or services and to let or request concessions pitches. The *World's Fair* is the single most important published source about fairgrounds and amusement parks during the twentieth century.
2  'White City Wonders', *World's Fair* (25 April 1908), p. 5.
3  'At the Franco–British Exhibition', *The Times* (9 June 1908), p. 8.
4  Javier Pes, 'Kiralfy, Imre (1845–1919)', *Oxford Dictionary of National Biography* (Oxford: Oxford University Press, 2004): http://www.oxforddnb.com/view/article/53347 accessed 11 January 2013.
5  'Open-air Pleasures in London', *The Times* (24 May 1909), p. 13. This tendency had been observed at the Crystal Palace American Exhibition in 1902 where a ride from

Coney Island, Loop-the-loops, outshone the manufacturing exhibits. Although the amusements did not yet amount to a coherent amusement park, the reporter noted that: 'it is not for exhibitions that visitors go to Crystal Palace. They go to enjoy themselves': 'American Exhibition at Crystal Palace', *The Times* (2 June 1902), p. 13.

6   'Visit of the Queen to White City', *London Daily Telegraph* (15 July 1909), sourced from the William Bean Scrapbook, Blackpool Pleasure Beach Archive (hereafter cited as BPBA).

7   See John F. Kasson, *Amusing the Million: Coney Island at the Turn of the Century* (New York: Hill and Wang, 1978); John K. Walton, 'Popular Playgrounds: Blackpool and Coney Island, c.1880–1970', *Manchester Region History Review* 17, 1 (2004), p. 52. By 1906, over 1500 parks operated across the US – see 'Park Notes', *Billboard* (3 February 1906), p. 20, quoted in Lauren Rabinovitz, *For the Love of Pleasure: Women, Movies and Culture in Turn-of-the-Century Chicago* (London: Rutgers University Press, 1998), p. 139.

8   This figure is based on a survey of parks featured in *World's Fair* from 1906 to 1939, and on the comprehensive lists made by Robert E. Preedy, *Roller Coasters: Their Amazing History* (Leeds: Robert E. Preedy, 1992), and *Roller Coaster: Shake, Rattle and Roll!* (Leeds: Robert E. Preedy, 1996). Reliable visitor statistics are scarce, but a sense of numbers can be gleaned from newspaper reports and other contemporary sources.

9   Peter Bennett, *Blackpool Pleasure Beach: A Century of Fun* (Blackpool: Blackpool Pleasure Beach, 1996), p. 18.

10   Walton, 'Popular Playgrounds', p. 54.

11   Gary S. Cross and John K. Walton, *The Playful Crowd: Pleasure Places in the Twentieth Century* (New York: Columbia University Press, 2005), p. 47.

12   At Blackpool Pleasure Beach, for example, gambling and gypsies were banned and the grounds were 'policed in accordance with the requirements of the Chief Constable': *Blackpool Gazette News* (12 April 1907), BPBA.

13   See J. Meredith Neil, 'The Rollercoaster: Architectural Symbol and Sign', *Journal of Popular Culture* 15, 1 (1981), pp. 108–15.

14   'Tober' is a term used to describe the site occupied by the fair.

15   Alexander Chase-Levenson, 'Annihilating Time and Space: Ecclecticism and Virtual Tourism at the Sydenham Crystal Palace', *Nineteenth-Century Contexts*, 34, 5 (December 2012), pp. 461–75.

16   On this theme see Brenda Brown, 'Landscapes of Theme Park Rides: Media, Modes, Messages', in Terence Young and Robert Riley (eds), *Theme Park Landscapes: Antecedents and Variations* (Washington: Dumbarton Oaks, 2002), p. 241.

17   The phrase 'gear and girder' was coined by the literary critic Cecelia Tichi to describe the pervasive impact of technology on American culture and aesthetics during the late nineteenth and early twentieth centuries: Cecelia Tichi, *Shifting Gears: Technology, Literature, Culture in Modernist America* (Chapel Hill: University of North Carolina, 1987), p. xiii.

18   Lieven de Cauter, 'The Panoramic Ecstasy: On World Exhibitions and the Disintergration of Experience', *Theory, Culture and Society* 10, 4 (1993), p. 12.

19   *World's Fair* (15 February 1908), p. 1.

20   David E. Nye, *American Technological Sublime* (Cambridge, Mass/London: MIT Press, 1994), p. xviii.

21   'New Water Chute', *Blackpool Times* (16 March 1907), BPBA.

22   Nye, *American Technological Sublime*, pp. 8–9. Presenting the sublime as cultural practice, rather than as an immutable law of perception (Edmund Burke), Nye describes the history of popular 'enthusiasms' for technological objects in the United States. Burke's *Philosophical Enquiry into the Origin of Our Ideas of the Sublime and Beautiful* was first published in 1757. Though Nye charts the development of what he calls a 'popular sublime', he draws heavily on the definition proposed by Burke (astonishment mingled with terror), and later developed by Kant (arithmetical, and dynamic sublime).

23  See, for example, John Henry Iles's Scenic Railway at White City in 1908, modelled on the Canadian Rockies: Preedy, *Roller Coasters*, p. 31; Jeffrey T. Schnapp, 'Crash (Speed as Engine of Individuation)', *Modernism/Modernity* 6, 1 (1999), p. 29.

24  *Souvenir of the White City* (1909), p. 23.

25  Nye, *American Technological Sublime*, p. 151.

26  'Our Al Fresco Entertainments', *Blackpool Herald* (23 July 1906), BPBA.

27  'The White City', *World's Fair* (2 March 1907), p. 1.

28  Rem Koolhaas, *Delirious New York: A Retroactive Manifesto for Manhattan* (New York: Monacelli Press, 1994), pp. 35, 41.

29  Blackpool Pleasure Beach Souvenir Booklet (c. 1912), Blackpool Central Library.

30  Rabinovitz, *For the Love of Pleasure*, p. 139.

31  'Kinaesthetic' is used here to describe the aesthetics of movement and multi-sensory modes of perception which were produced and experienced at the amusement park.

32  Wolfgang Schivelbusch, *The Railway Journey: The Industrialization of Time and Space in the 19th Century* (Berkeley and Los Angeles: University of California Press, 1986), p. 194.

33  Ibid., p. 189.

34  Ibid., p. 193.

35  Ibid., p. 191.

36  'A Gay Time on Blackpool Pleasure Beach', *Blackpool Times* (7 September 1907), BPBA.

37  Ibid.

38  For a detailed account of early ambivalence towards railway travel see Schivelbusch, *The Railway Journey*, pp. 5– 15.

39  Ibid., p. 159.

40  Georg Simmel's 'The Metropolis and Mental Life' was originally published as 'Die Grosstadt und das Geistesleben' (1903), trans. reprinted in Donald N. Levine (ed.), *On Indiviuality and Social Forms* (Chicago/London: University of Chicago Press, 1971), p. 329. See also David Frisby, *Fragments of Modernity: Theories of Modernity in the Work of Simmel, Kracauer and Benjamin* (Cambridge, Mass/London: MIT Press, 1986), p. 73–4.

41  Sigmund Freud, *Beyond the Pleasure Principle, 4,* James Strachey (ed.) (London: Hogarth Press and Institute of Psychoanalysis, 1974), pp. 20–22.

42  Frisby, *Fragments of Modernity*, p. 75; Bennett, *Pleasure Beach*, p. 14.

43  Frisby, *Fragments of Modernity*, p. 74.

44  Georg Simmel, *The Philosophy of Money* (1907), trans. by Tom Bottomore and David Frisby (London/Boston: Routledge, 1978), p. 257; cited by Frisby, *Fragments of Modernity*, p. 74.

45  Simmel, *The Philosophy of Money*, p. 484; cited by Frisby, *Fragments of Modernity*, p. 72.

46  'The Mad Rush for Pleasure', *World's Fair* (15 January 1927), p. 21.

47  'Rainbow Pleasure Wheel', *World's Fair* (24 February 1912), p. 8.

48  'All About the Mammoth Fun City', *World's Fair* (31 August 1907), p. 6.

49  The comfort and enclosure of late-nineteenth century trains repressed fears of accidents and danger: Schivelbusch, *The Railway Journey*, p. 162.

50  'All About the Mammoth Fun City', *World's Fair* (31 August 1907), p. 6.

51  'A Fortune in a Thrill', *The Sunday Chronicle* (Manchester, 23 August 1908), BPBA.

52  Arwen Mohun, 'Design for Thrills and Safety: Amusement Parks and the Commodification of Risk, 1880–1929', *Journal of Design History* 14, 4 (2001), p. 292.

53  'Figure 8 Railway Accident at Cleethorpes', *World's Fair* (4 June 1910), p. 7.

54  *Royal Society for Prevention of Accidents*: http://www.rospa.com/about/history/ accessed 26 April 2015. Formalised attitudes towards safety came surprisingly late in the twentieth century with the 1937 Factory Act.

55   'Scenic Railway Accident', *World's Fair* (19 August 1911), p. 12.
56   Two workers were seriously injured during the construction of the Pleasure Beach's Scenic Railway in May 1907 – one from a fall, the other was electrocuted: 'Fairground Accident', *Blackpool Herald* (24 May 1907); 'Fell Thirty Feet', *Blackpool Herald* (10 May 1907), BPBA.
57   Mohun, 'Design for Thrills and Safety', pp. 292, 300.
58   'A Day in Breezy Blackpool. A Pleasure Mill by the Sea', *London Standard* (24 August 1906), BPBA.
59   Mohun, 'Design for Thrills and Safety', p. 294.
60   'Canadian Sketches', *Manchester City News* (8 September 1906), BPBA.
61   Pleasure Beach Souvenir Brochure (c. 1912), Blackpool Library Collection.
62   Lucy Fischer, '"The Shock of the New": Electrification, Illumination, Urbanization, and the Cinema', in Murray Pomerance (ed.), *Cinema and Modernity* (New Brunswick/London: Rutgers University Press, 2006), p. 36.
63   Maurice Elvey, *Hindle Wakes* (Gaumont British Picture Corporation, 1927).
64   Bennett, *Pleasure Beach*, p. 20.
65   In 1884, Sir Hiram formed the Maxim Gun Company. His fully automatic gun was adopted by the British Army in 1889, and the Royal Navy in 1892. The company was later absorbed into Vickers Sons and Maxim. Variants of Maxim's machine guns were used by British, German and Russian troops in both World Wars: Brysson Cunningham, *Maxim, Sir Hiram Stevens (1840–1916), Oxford Dictionary of National Biography* (Oxford: Oxford University Press, 2004): http://www.oxforddnb.com/view/article/34954; Arthur Hawkey, *The Amazing Hiram Maxim: An Intimate Biography* (Staplehurst: Spellmount, 2001), p. 104.
66   'South Shore's Newest Novelty', *Blackpool Herald* (22 February 1907), BPBA.
67   David Edgerton, *England and the Aeroplane: An Essay on a Militant and Technological Nation* (London: Macmillan Press, 1991), p. 2; Robert Wohl, *A Passion For Wings: Aviation and the Western Imagination* (London: Yale University Press, 1994), p. 14.
68   *Souvenir of the White City* (1909), p. 24, Manchester Room and County Record Office, acc. 791 M24.
69   'On the Pleasure Beach, by a Visitor', *Blackpool Times* (24 August, 1907), BPBA.
70   'Novelties for the Mammoth Fun City at Olympia', *World's Fair* (14 December 1907), p. 6 (quotes extensively from a recent edition of the *Daily Express* national newspaper).
71   *Submarine*, Encyclopaedia Britannica Online Academic Edition: http://academic.eb.com.ezproxy.westminster.ac.uk/EBchecked/topic/570813/submarine; Bernhard Rieger, *Technology and the Culture of Modernity in Britain and Germany, 1890–1945* (Cambridge: Cambridge University Press, 2005), p. 227.
72   Daryl Adair, 'Spectacles of Speed and Endurance: The Formative Years of Motor Racing in Europe', in David Thoms, Len Holden and Tim Claydon (eds), *The Motor Car and Popular Culture in the Twentieth Century* (Aldershot: Ashgate, 1998), p. 129.
73   'Motor Car Racing', *Blackpool Gazette News* (28 August 1908), BPBA.
74   'A Safe Brooklands', *Daily Express* (12 September 1908), BPBA.
75   'Motor Car Racing', *Blackpool Gazette News* (28 August 1908), BPBA.
76   'The World's Fairograph', *World's Fair* (29 December 1906), p. 4.
77   'The Rolling Motor Track', *World's Fair* (30 July 1910), p. 5.
78   The RAC initially refused women members, and in defiance of popular disapproval a Ladies Automobile Club was formed in 1903: Sean O'Connell, *The Car in British Society: Class, Gender and Motoring 1896–1939* (Manchester: Manchester University Press, 1998), p. 48.
79   Thomas Misa, 'The Compelling Tangle of Modernity and Technology', in Thomas Misa, Philip Brey and Andrew Feenberg (eds), *Modernity and Technology* (Cambridge, Mass/London: MIT Press, 2003), p. 12.

80   'White City Wonders', *World's Fair* (25 April 1908), p. 5.
81   'Fairground Novelties', *Blackpool Times* (20 February 1907), BPBA.
82   'The Latest Novelty', *Blackpool Herald*, (24 May 1907), BPBA.
83   Rieger, *Technology and the Culture of Modernity in Britain and Germany*, p. 16.
84   Ibid., p. 224.
85   *Souvenir of the White City* (1909), p. 30, Manchester Room and County Record Office, acc. 791 M24.
86   'Bank Holiday', *Blackpool Herald* (6 August 1907), BPBA.
87   'On the Pleasure Beach' *Blackpool Times* (24 August 1907), BPBA.
88   'Bank Holiday . . . At The Franco–British Exhibition', *The Times* (9 June 1908), p. 8.
89   'Blackpool: On the Seafront at Holiday Time', *Lancashire Daily Post* (16 August 1907).
90   New critical discourses emerge in the 1920s and 30s that depicted the mechanisation of commercial entertainments as the cause and symbol of working-class degeneration. For an examination of the reactions of the British educated elite to the development of commercial gramophone, radio and cinema culture see Dan LeMahieu, *A Culture for Democracy: Mass Communication and the Cultivated Mind in Britain Between the Wars* (Oxford: Clarendon Press, 1988). These ideas have continued to influence elitist attitudes toward all manner of entertainments – from amusement parks and cinema, to the juke box and computer games.
91   'What Will Southport Do?', *Blackpool Herald* (19 February 1907), BPBA.
92   'On the Pleasure Beach', *Blackpool Times* (24 August 1907), BPBA.
93   Ken Crowe, *Kursaal Memories: A History of Southend's Amusement Park* (St Albans: Skelter Publishing, 2003), p. 15.
94   'Blackpool's Newest Entertainment', *World's Fair* (20 August 1910), p. 14.
95   'On the Pleasure Beach', *Blackpool Times* (24 August 1907), BPBA.
96   'Mrs Ashley and Party: Round the Pleasure Beach', *Blackpool Gazette News* (23 July 1909), BPBA.
97   'South Shore Pleasure Beach', *Blackpool Times* (30 March 1907), BPBA.
98   'Blackpool's Attractions', *Oldham Chronicle* (15 August 1908), BPBA.
99   'London to Have its Blackpool', *World's Fair* (4 May 1907), p. 4.
100  'Opening of the Edinburgh Marine Gardens', *World's Fair* (7 May 1910), p. 9.
101  From the 1830s, the predominantly middle-class rational recreation movement campaigned for organised and edifying non-work activities for the working classes: Chris Rojek, *Ways of Escape: Modern Transformations in Leisure and Travel* (Basingstoke/London: Macmillan Press, 1993), pp. 32–4.
102  *Blackpool Gazette News* (19 March 1907), BPBA.
103  'The Fall from a Slide at Blackpool', *World's Fair* (2 June 1906), p. 4.
104  'Old Folks Become Young Again', *Blackpool Gazette News* (16 September 1911), BPBA.
105  'A White City Accident', *The Times* (10 May 1910), p. 12.
106  'The Taste for "Thrills"', *World's Fair* (17 August 1912), front page.
107  Simmel, 'The Metropolis and Mental Life', pp. 324–40.
108  Ibid., p. 325.
109  Frisby, *Fragments of Modernity*, pp. 73–4.
110  'Blackpool's Carnival of Sensations', *The Sunday Chronicle* (Manchester, 31 July 1910), p. 2.
111  Cross and Walton, *The Playful Crowd*, pp. 4–5, 61–2, identify the 'playful crowd' at the amusement park, a phenomenon which, they argue, represented a modernised expression of pre-industrial saturnalia. Elements of medieval carnival surviving at Blackpool Pleasure Beach are identified by Tony Bennett, 'Hegemony, Ideology, Pleasure: Blackpool', in Tony Bennett, Colin Mercer and Janet Woollacott (eds), *Popular Culture and Social Relations* (Milton Keynes: Open University Press, 1986), pp. 135–54.
112  My understanding of the carnivalesque is based on a reading of Mikhail Bakhtin, *Rabelais and His World*, trans. Hélène Iswolsky (Bloomington: Indiana University Press, 1984).

113   'The White City Make Another Application for a Drink License', *World's Fair* (15 February 1908), p. 5.

114   '"White City" License Granted', *The Times* (2 May 1913), p. 2.

115   Rob Shields locates the carnivalesque on the beaches of Brighton and Margate: Rob Shields, *Places on the Margin: Alternative Geographies of Modernity* (London: Routledge, 1991). Blackpool's Golden Mile, with its wax works and freak shows, offered visitors graphic inversions of social norms: see Gary Cross, 'Crowds and Leisure', *Journal of Social History*, 39, 3 (Spring 2006), pp. 635–6.

116   'The Morals of Blackpool', *John Bull* (1 May 1909), BPBA.

117   Bakhtin, *Rabelais and His World*, pp. 6, 9.

118   Ibid., p. 11.

119   Ibid., p. 7.

120   'Blackpool's Carnival of Sensations', *The Sunday Chronicle* (Manchester, 31 July 1910), p. 2.

121   Ibid.

122   Kasson, *Amusing the Million*, p. 109.

123   Darren Webb, 'Bakhtin at the Seaside: Utopia, Modernity and the Carnivalesque', *Theory, Culture and Society* 22, 3 (2005), pp. 121–38.

124   Ibid.

125   Ibid., p. 133.

126   Ibid.

127   'Blackpool's Carnival of Sensations', *The Sunday Chronicle* (Manchester, 31 July 1910), p. 2.

# 3   From Battersea to Alton Towers

## In search of the Great British theme park

*Ian Trowell*

In 1951, Battersea Park in London was transformed into the 'Festival Gardens' as part of the Festival of Britain celebrations. This transformation included the addition of amusement park attractions forming what became known as Battersea Amusement Park. Together, these new attractions created a vibrant point in the city of London. Although the Festival only had a year-long schedule, the amusement park continued through until the onset of the 1970s as a fine example of a city-based amusement park, equivalent to other British locations such as Belle Vue in Manchester and Sutton Coldfield in Birmingham. Following a fatal rollercoaster accident in 1972, the park went into a rapid decline, with total closure completed by the end of the decade. Its eradication, both physically and as a cultural memory, was so swift and complete that the next generation of amusement spaces emerging at the start of the subsequent decade – pioneered by a significantly upgraded Alton Towers – were built and marketed with little or no reference to earlier ventures like Battersea.

In recent publications cited throughout this chapter, the term 'theme park' is applied loosely to the Festival of Britain and its amusement spaces,[1] with an acknowledgement that perhaps this was a theme park before the term came in to use. Equally, there is a commonly held opinion amongst amusement park enthusiasts that Alton Towers was the start of the contemporary theme park movement in the UK, with its investment of up-scaled mechanical attractions, even though when pressed it is difficult to determine what exactly the 'theme' of Alton Towers purports to be.

The definition of theme park has proved and continues to be both nebulous and amorphous, extending as a concept that reaches beyond into other social and cultural spheres – shopping malls, lavish ocean cruise ships, and in the packaging of events such as pop concerts or football matches. As Scott Lukas has remarked, in his attempt to disentangle history and definitions, a contemporary amusement space is not necessarily a theme park and contemporary non-amusement spaces can equally be understood as theme parks.[2] What emerges is an approximate model whereby the first amusement space assembled by Walt Disney in 1955 at Anaheim can be considered as both a clear development and rupture of previously existing amusement spaces. Disney, in turn, developed the model of a theme park as an aggressive concept such that the take-up of the model of the theme park was based upon an advanced model that had undergone numerous iterations.

This chapter takes a British focus and draws on the rich debate in establishing models and approaches towards the understanding of amusement spaces throughout the twentieth century. Firstly, the question as to what constitutes the first British theme park and why – if indeed Britain ever had a theme park – will be considered through the two axis points of Battersea Amusement Park (and its realisation through the Festival of Britain) and the 1980s re-branding of Alton Towers. Secondly, the development of a new model is put forward to consider the theme park and to look at current spaces of amusement that have taken the success of Alton Towers as their blueprint.

## The Festival of Britain – A first theme park?

By the early 1950s, Britain maintained a healthy surplus of amusement spaces. Seaside amusement parks existed in nearly every sea-bordering county in England, Scotland and Wales – as an example, the north-east England had parks at Whitley Bay, South Shields, Seaburn (Roker), Seaton Carew (Hartlepool), Redcar and Saltburn crammed onto a stretch of coast of only about 30 miles in distance. The traditional travelling fairground had re-established itself after the war and the UK was enjoying a large number of events. As well as Belle Vue and Sutton Coldfield, smaller inland amusement parks thrived in parks and gardens at Alton Towers, Drayton Manor, Wicksteed and Burnham Beeches. The tradition of expositions and world fairs had understandably stalled with the Glasgow Empire Exhibition of 1938 falling just before the outbreak of hostilities; however, the Festival of Britain would re-address this perceived imbalance.

The theme park concept is considered as a hybridisation of these existing forms of amusement spaces with an added ingredient that seemingly attains the mystical aura of the recipe for Coca-Cola. Within these spaces there further exists intertwined forms of entertainment from technological imperatives, appreciation of spaces, performance and exhibition. The study of these spaces and forms has created a complex debate based around various approaches in regard to the nature of the enquiry (the subjective view of the 'punter' set against the subjectifying view of the amusement proprietor) combined with theoretical tools adopted from sociological and political schools of thought. Attempts to work with an overarching approach – either chronological or teleological – and to create a genealogy of types can run aground at the first instance or encourage a slew of counter-responses.[3]

None of these overarching histories make reference to the Festival of Britain/Battersea, though this can be assumed to stem from their American bias where the growth of Coney Island as an amusement park nexus, the 'midway' fairs and the White City Columbian Exposition of 1893 can be seen as the raw ingredients that go towards a possible recipe for 1955's seminal Disneyland Park. Certainly, laid out in a simple chronological timeline, the 1951 Festival of Britain can be seen as the closest relative (the missing link?) in the exposition/world's fair format to the Disney phenomenon.

In the first instance, the wider Festival of Britain encompassed the themed building and provision of amusements at Battersea. The technological imperative that defined the amusement device – later to play a key role in determining the social conception of the theme park – was secondary in this case to the technology on show throughout the Festival. Technology here can be grasped in the context of exhibition culture. The exposition often saw these elements combined into a refined and well-rehearsed complete package and this polished form provides another link to the emergent theme park.[4]

The planning of the Festival of Britain began shortly after the end of the Second World War, and had various intentions to act as a 'tonic to the nation'. This included 'a pat on the back' for the war effort, reaffirmation of the country's contribution to world civilisation, and a democratic engagement of the arts, science and technology. The proposed date of 1951 (between May and September of that year) was a direct reference to the centenary of the 1851 Crystal Palace Exhibition – seen as the first occurrence of these 'ephemeral vistas'. It was also utilised as a 'soon as possible' date since the agreed sequence of official world's fairs would not permit a vacancy until much later in the 1950s.[5] Such a deliberate and structured attempt to integrate a sense of 'celebration' with vision under a political rubric was obviously not without contention and internecine controversy within the political strata. The mood of austerity that endured in the post-war years was not generally seen to be fitting to a large-scale celebration, and social and cultural commentators of the time found plenty of ammunition to wage a war of discontent.[6] This disillusionment had persisted through into critical considerations of the whole event with debates around the Festival being a tool of the government and an oppressive 'taste making' exercise. Elsewhere, the initiative to present Britain as tourist destination via the Festival floundered equally on ideological rocky ground.[7]

Battersea Festival Gardens and the amusement park grew almost as an adjunct to the main Festival – classed in some quarters as a poor relation.[8] As with previous expositions and exhibitions the inclusion of an amusement space complete with fun fair elements caused some tension but the general feeling was that events such as these needed such a space as an antidote to the seriousness of the occasion. Some provision towards relief and frivolity had already been accommodated within the Lion and Unicorn Pavilion, though an amusement park within a themed space was proposed on a self-financing basis and allowed the Battersea site to gain a five-year lease in an attempt to recoup outlay costs.[9] The Festival Gardens were laid out on the site of allotments and a cricket pitch, with design and theming overseen by James Gardner. The spirit and importance of Vauxhall Gardens was evoked, with the intention to relax and have fun – elegant fun.

The larger space of the Festival Gardens included a zoo and pets section, theatres, a tree-top walk and large dance tent pavilion. The core amusement park itself was not overtly themed – this would have been difficult as the 'theme' of the Festival of Britain was design from past, present and future. With this in mind, there is the suggestion of some controversy in the fair with the inclusion of American amusement rides. It was not uncommon for seaside amusement parks to invest in occasional American rides – these being not necessarily more advanced than British rides but differently advanced. The appointed fairground operatives at Battersea embarked on an expenses-paid trip to scope out some American rides and the opening of the fair saw a total of four such examples. Becky Conekin's history of the Festival of Britain asserts a theory about the American amusement park rides at Battersea, in line with her general chronology of tensions and controversies but it would be foolish to consider the fair as an American import. A reading of the weekly *World's Fair* newspaper up to and around the opening of the fair gives a better balance as to the heritage and pedigree of the fairground machines on offer. The American thrills were the Bubble Bounce (a novelty but not a success), the Fly-o-plane Jets (again a novelty that was not taken up again in the UK), the Hurricane (a strange machine that was soon replaced) and something called the Flying Cars (no pictures of this have survived). The major attraction was John Collins' Rollercoaster (brought down from the amusement park at Sutton Coldfield), a fantastic Rotor (designed in collaboration with inventor Ernst Hoffmeister and Burton-on-Trent manufacturers Orton and Spooner), and a large selection of British fairground rides taken off the travelling circuit (Figures 3.1 and 3.2). Later structures included the Water Chute and the famous

*Figure 3.1* Entrance to the Battersea Festival Gardens and amusement park

Source: © National Fairground Archive, University of Sheffield

*Figure 3.2* General view of Battersea Festival Gardens and amusement park

Source: © National Fairground Archive, University of Sheffield

Sky Wheels (both built in the UK) (Figure 3.3). Control and organisation of the space had passed to core people already involved in running amusement parks and fun fairs – notably Sir Leslie Joseph of park fame and Granville Hill of the Showmen's Guild of Great Britain – and the amusement park functioned in a standard fashion by interchanging attractions with various travelling fairground machines.[10] As the lease was extended further, Charles Forte became prominent in directing the arrangements, continuing a growing relationship between Forte and amusement provision in the UK.

The park attained a steady dynamic housing the latest attractions as prominent concessionaires like the Botton Brothers took up involvement. Battersea Amusement Park thrived as a cultural focal point in central London as the 'swinging sixties' took hold as a wider theme for the city itself; the park featuring in films such as *Wrong Arm of the Law* (1963) and pop promotions such as Cliff Richards' *Flying Machine*. However, throughout the decade alternative venues of amusement were growing around the London hub and investing in amusement park technologies – Chessington Zoo, Thorpe Park (originally a flooded gravel pit with boating attractions), Woburn Abbey, Windsor Safari Park, Whipsnade Zoo – and this competition marked a gradual decline in Battersea.

With the end of the extended lease in sight there were plans to revamp the operation towards what would have been the first UK (post-Disney) theme park. This ambitious plan was curtailed when a major accident occurred on the original

*Figure 3.3* Battersea amusement park's pioneering Sky Wheels

Source: © National Fairground Archive, University of Sheffield

rollercoaster in 1972. The park briefly invested in a modern new Pinfari steel coaster and plans were again mooted for a Disney-style theme park in 1974 – the 'Magic World' but none of this came to light.

Battersea Amusement Park was thus initially part of the re-branded Battersea Festival Gardens, which themselves were part of the much larger Festival of Britain located ostensibly on and around the heavily re-themed spaces of London's South Bank. The input of the exposition tradition into the genesis of Disney's first park in 1955 is an accepted argument, and by its proximity in time the Festival of Britain must be considered as a concept close to a theme park as is possible. Moreover, the theme of the Festival of Britain had a greater focus than previously occurring world's fairs and expositions due to the circumstances and timing after the Second World War. The planning of a visual identity is a key part of the event and shows evidence of the move towards a tightly branded, interlocking series of spaces.[11] The emergence of themed zones and a constructed and consistent architecture to maximise an escapist and immersive experience in the service of display and exhibition is a key driver for the theme park. Lukas echoes previous work as he identifies the World's Columbian Exhibition of 1893 giving 'a powerful glimpse of what a theme park could be'.[12] The designed space moves towards a Gesamtkunstwerk[13] and attains an overarching identity in and of itself, awash with rigorous symbolism. The architecture sets out to enclose, embellish and foster a new sense of place.

Considering Battersea Amusement Park as a theme park in its extended life is, however, not a natural corollary of understanding the Festival of Britain as a prototype themed space of amusement. Rather, it can be suggested that vestigial influences and linkages persisted in the perception of the space, and certainly the re-surfacing of 'Britishness' as a tangible thing alongside the swinging sixties that helped to re-boot Battersea Amusement Park as something to be experienced in a vaguely themed manner. A further crucial ingredient that cemented the theme park as something different is missing here; namely the negotiations to assert extended behavioural control through themed space. Conekin is keen to empha-sise the tensions evident from the start of the Festival of Britain:

> the traditional view of the fair as the definitive unruly and dangerous place of crowds, disorder and distasteful pleasures haunted the middle-class planners of the Festival Gardens who repeatedly stressed that they did not want this place to be known as a fair.[14]

It was a general view cast onto most spaces of mechanised amusement as evi-denced by J. B. Priestley who, in his survey of the British public, expressed similar horror at the general idea of the amusement.[15] Disney dreamt up spaces of amusement that would naturally and instinctively banish such displays of revelry.

Another crucial difference between the Festival of Britain and Disney's first theme park is sited on the cultural time-line marking Disney's close relationship with the constructed reality of film and television. Perhaps the most incisive obser-vation is made by Paul Rennie who assesses the idea of the Festival as occurring at a 'decisive moment in mass culture . . . the last event to be produced in a cul-tural landscape not dominated by television'.[16] This observation will be returned to towards the conclusion of the chapter.

## Alton Towers – The machine and the garden

A permanent site for the provision of fairground amusements in London ended with Battersea in 1974. The capital was, however, flush with many large and established travelling fairs, and day-trippers looking for the tradition of the amusement park did not have far to go to reach Southend's Kursaal or Margate's Dreamland. The climate was changing though, and as the 1980s progressed the wealth of amusement parks around British seaside resorts began to decline rapidly (see Nick Laister, this volume). The city-based parks at Manchester (Belle Vue) and Birmingham (Sutton Coldfield) were also nearing the end of their existence as the country geared itself for something different. This change occurred in 1980 when Alton Towers shook up the industry with the installation of the Vekoma

Corkscrew rollercoaster. A quarter of a century after the Disney's 1955 theme park, the UK now had what might be considered its first. Alton Towers, therefore, as something new and spectacular, presents an instructive case for examination.

To the generation of thrill-seekers brought up in the UK in an era of instant communication and digital experience sharing there is an overriding presumption that Alton Towers as an amusement space simply did not exist prior to the installation of the Corkscrew.[17] Clearly, this is not the case, and the provision of amusement spaces at Alton Towers has a rich history that mines an earlier vein of 'Britishness'.[18] Opened to the public in 1839 as a leisure garden, Alton Towers was a typically class-delineated space for those wishing to admire its features. Ironically, these consisted of what could be described as Victorian theme park elements – Chinese pagoda, gothic monuments, a Swiss cottage. Following a move to private business ownership, the park engaged a more commercial strategy to bring it into line with similar spaces in twentieth-century Britain. A zoo, amusements and caravan site were added through the years as the park trod a similar path to other pleasure gardens and zoos such as Chessington, Drayton Manor, Flamingo Land, Trentham Gardens, Dudley, etc. (Figure 3.4). These spaces of enjoyment and relaxation were competing for commercial viability in a shrinking market, as the British public began spending spare money on the first of the cheaper package holidays into Europe. Such inland, verdure attractions were also in competition with UK seaside resorts, and had to fight hard to give the paying public that little bit extra on top of 'getting away from it' for the day. Consequently, Alton Towers worked hard to develop commercial link-ups with national newspapers such as the *Daily Mirror*.

*Figure 3.4* Traditional amusements amongst the pleasure gardens of Alton Towers, 1968

At the end of the 1970s, it would appear that a radical decision had been reached (though such intentions were not public knowledge). Even as late as the tail end of 1979, *World's Fair* ran a feature on the park through concessionaire Brian Collins who talked about his provision of standard fairground-style amusements within the park.[19] However, only six months later, the same newspaper ran a feature announcing the opening of the Corkscrew (Figure 3.5), the plans to come, and the fact that Brian Collins had now been re-assigned to open just a single amusement arcade on a one-year lease.[20] John Broome, with the management team, had conducted a detailed study of European and American theme parks and had deemed that traditional fairground rides were no longer adequate. Broome's intentions were both rooted in the success of Disney and part against its overarching plastic totality:

> Disney is a particularly American phenomenon which is hard to sell with a certain artificial element in it. It is a beautifully-run concept. The English market demands a little bit more of a solid approach and that is what we are doing here; we are putting these attractions within a real country garden setting.[21]

Broome wanted to emphasise the gardens and beauty of his site, whilst providing a unique and ultra-modern thrill. As the park commenced its fourth year of this new life, interviews with Broome had shifted ground when the Disney-word was mentioned – the Britishness that defined the park against a notion of manufactured Disney-fication had taken hold as a manufactured theme of Britishness to include

*Figure 3.5* The debut of the Corkscrew rollercoaster in 1980

Source: © National Fairground Archive, University of Sheffield a Victorian Street, ice cream parlour, tea 'shoppe', archaic photographic studio and a theatre that performed 'Rule Britannia' to enthusiastic crowds patriotically overblown with the recent Falklands conflict of 1982.[22]

New initiatives such as a paid entry fee to include all attractions, and a stream-lined approach to the management of people flow and emotional satisfaction, aligned Alton Towers with the Disney model. However, Broome had made it very clear he did not want to replicate the intensive plastic and concrete approach of Disney, and an obvious consequence of not doing this was thus forsaking the chance to foster a Disney-style 'mindset' within the guest. The theming of Disney – this double abstraction – to immerse the guest into a replication of a pre-constructed fantasy world, was not initially considered at Alton Towers. Instead, the vibrancy and splendour of the park and gardens was utilised, not necessarily as theme but as juxtaposition.

Though Alton Towers had made the bold move to step up a whole set of gear changes by investing in the next generation of amusement technology, it still traded on British traditions with the history of its gardens and the marketing of the mechanical behemoths of amusement through a type of fairground related spatial phenomenon. Early advertisements celebrating the heavy investment of Alton Towers into these theme park standard attractions played upon the juxta-position between tranquillity and forested serenity, and the groaning, creaking, overwhelming monstrosity of these massive thrill rides. Having visited the park as a teenager prior to 1980 and through the onset of the larger investments, this author is still struck by the memories of switching from leafy and coniferous con-fines to giant lurching structures in the blink of an eye (Figure 3.6).

Alton Towers had firmly set the agenda with the soaraway success of the Corkscrew, and fellow 'theme parks' such as Drayton Manor, Chessington and Thorpe Park were soon following the trend. A new type of inland theme park was seen as the place to experience the biggest and latest rollercoasters but it begs the question as to how these theme parks were 'themed' and experienced.

## Experiencing spaces of amusement parks

As mentioned earlier, the spaces of amusement parks have been the source of much debate in terms of approach and interpretation. They are heavily contested and mediated, and evolve through time both with particular environments and across into new environments with the (current) version of the theme park. The second part of this chapter takes a further look at the spaces where amusement technologies are experienced in an attempt to go beyond the axis points of the Festival of Britain and the re-investment in Alton Towers. For this, it is neces-sary to set out some groundwork for further approaches to analysing amusement spaces – principally to look at work around spaces and places (in general) and to see how technologically mediated amusements (fairground devices, thrill rides, etc.) have moved to the forefront of this type of environment.

For the purpose of looking at the spaces and places of amusement sites it is advocated that a space becomes a place when it is imbued with a sense of purpose,

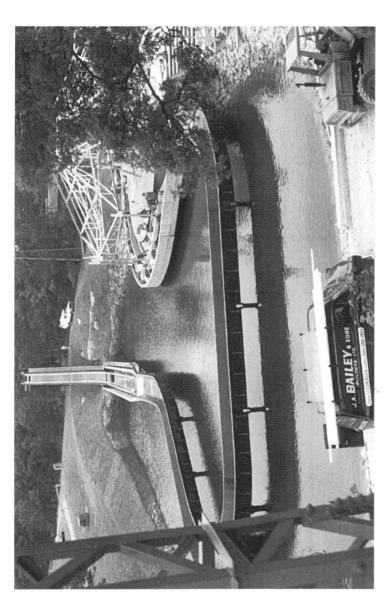

*Figure 3.6* The landscape becomes postmodern with the lakes of the Alton Towers Log Flume created on top of the pre-existing lakes of the pleasure gardens

structure and population at a certain point in time.[23] A place then creates a sense of certainty and routine for individuals and groups utilising or inhabiting the place. There are times when the smooth transition from space to place does not quite flow, and times when the bond between space and place starts to separate. A useful tool for thinking about these spatial relationships is Michel Foucault's proposal of heterotopic spaces, though in typical fashion Foucault does not set out to provide a jigsaw-like fit to bring a portion of metaphysics into conceptual order.[24]

Heterotopias are given as examples of types where function seems to be out of kilter with current time (cemeteries, mental hospitals) and Foucault picks out the fairground carnival as a specific example. In other situations, the dis-functioning of a place might be sudden, radical and temporary – imagine the feeling when, stuck stationary in a seemingly endless traffic jam on a scorched and barren motorway, you step out of your vehicle onto the road surface. Hence, a useful way of thinking about a heterotopia might then be a situation where the bond between a space and place becomes decoupled, and associated with this decoupling emerges the possibility of 'something else', potentially a sense of carnival. It is this chain of events that can be taken forward to look at a sequence of spaces that will allow an understanding of the notion of the modern amusement park nested in the normalised rubric of the theme park.

The travelling fairground is a good place to start. The fairground arrives overnight, extends a visit for a few days (or nights), and then leaves with barely a trace: it is thus both temporary and transient. The fairground needs a pre-existing space to occupy, and since the fairground needs to come into the orbit of a wide public, it will occupy a place (rather than an inert space). Two broad types of spatial arrangement are then constructed; either a bounded and compact region of amusements fitting into a single place (a green park, a car park or a large market place), or the fair will extend into a network of plots in a larger space and work around set structures (buildings, street layout, etc).[25]

The fairground, particularly the street fair, radically transforms the place that is previously defined within the well-grooved and routine vectors of normal existence – work, commuting, trade, shopping. It offers a brief but violent transformation of place and purpose. In transforming the space of everyday life, the fairground is raucous and boisterous; the unfolding of events at any particular time is highly unscripted. The Bakhtinian potential of radical carnival – a possibility of a number of effects such as suspension of hierarchy, prohibiting the norms of routinised time, dissolving of the individual's sense of self – is picked up as a critical point.[26]

Seaside amusement parks would be the next space/place to consider but it is arguable that the seaside amusement park was actually a space within a place. The amusement park sits within the particular seaside environment, and the seaside is clearly experienced as a place on a number of levels. The amusement park itself might also be considered a further place within the experience of the seaside but the 'playful crowd'[27] is firstly adapting to the seaside and secondly adapting to the spaces of particular amusements within the seaside itself. This approach contrasts with the ideas developed by Josephine Kane who uses Simmel's focus on the crowd to particularise the amusement park as an environment to satisfy desires for extrapolating the things that characterise modernity (technology, machinery, thrill, movement). Kane thus focuses on what is experienced in the amusement space rather than how the space itself (whether the amusement park or the wider seaside environment) is experienced.[28]

Gary S. Cross and John K. Walton's playful crowd is not as easily defined and malleable as Tony Bennett's crowd, ripe with Bakhtinian potential and under constant hegemonic management and manipulation by the powers that be. This debate can get entrenched in the potential of the crowd, what they 'should' be doing and perhaps why they are not doing it – self-policing, etc.[29] Thinking outside of this, there is potential to understand the experience of the seaside as something verging on the heterotopic. The visit to the seaside does not attempt to eradicate traces of everyday life; in some regards the reverse is true whereby it instead acknowledges that the drudgery of the every day is always there and can at best be put to one side. It plays upon the encroaching return to normality with jokes and marketing phrases urging you to savour the moment and 'kiss me quick'. The tangible structure of what turns the seaside space into a place (for the day-tripper) is only appreciated in brief passing, and it is the nature of this brief passing that in turn creates the strangeness of the character of the seaside place. For example, a seaside place experienced squarely out of season, in the depths of winter, is an unsettling spatial and temporal experience. Visits are only made in a manner akin to trying to find out if the refrigerator light does truly go out when you shut the door.

Outside of this relationship to the place forged through the constraints of transient visitation there is also a sense of marginality and liminality associated with the seaside resort. The seaside is not just a limit of marginality in terms of moving undesirable behaviour onwards and outwards until it reaches a geographical impasse (the sea itself); it is a strange, heterotopic site in itself with the land, the beach and shingle, and the sea in quick succession. Rob Shields has explored these 'places on the margin' and remarks on discontinuities in social fabric, loss of coordinates and moments of inbetweenness.[30]

The sampling of amusements with the exposition is more complex to relate as a place, and the aforementioned non-spatial approaches of both Kane (the crowd hungering for unfulfilled modernity) and Bennett (the reigning in of disordered and dangerous knowledge) can be applied more readily here. Expositions worked with a barrage of temporal stimuli as opposed to spatial – exhibits from the past, present and (possible) future(s) were mixed to give a poly-temporal extravaganza or 'an overlap of temporal registers'.[31] These expositions were not necessarily run in collaboration with showmen and the inclusion of amusements often created tension as these events established themselves into the twentieth century. Early expositions included gargantuan structures such as wheels, towers and switchback railways, these dovetailing with the rationale of expressing the various spirits of culture, innovation, character and industry. The planning and motivation for these events stemmed from governmental and national directives, and so those attending were expected to be imbued with some form of education, awe and wonder. The later tension between amusements within the exposition and the wider conceptual aim of the event led to an architectural and planning approach such that the amusement area was often distinct from the whole event. Expositions were crucial in developing these sectored and themed spaces, though it can be suggested that the amusement park was not prioritised in being considered as an 'interlocking sector' nor necessarily as a continuously themed concept – it was more of an adjunct. The overall spaces utilised by expositions were generally not previously associated as places so an element of re-purposing was not a factor (the Festival of Britain utilised ostensibly bomb-damaged areas on the South Bank), and the exposition provides a further link here to the theme park: we can see here the origins of 'placelessness' that characterise the emerging post-1950s theme parks.

Disney's first theme park at Anaheim developed pre-existing ideas and added new twists in equal measure, a strategy remarked by Judith Adams: 'On the 17 July 1955 an American icon with perhaps no match in popular culture exploded into being amid the orange groves of Anaheim, California, irrevocably altering the course of the amusement industry'.[32] Whereas continuity can be stressed, rupture and revolution are also significant. It is arguable that the two key factors that distinguished Disney's venture as a break from previous spaces are the interlocking aspects of deliberate placelessness and re-mediation of experience.

Disney wanted to tame the playful crowd away from the disorder and unscheduled sensuality that he had witnessed at Coney Island, marking a conceptual shift in the design and organisation of the amusement space. There was a move from simply giving the public what they wanted towards considering what the operators of the theme park might want in the process of giving the public what they want. Key in this was the creation of a placeless space,[33] primarily gained through building on unconsidered agricultural areas and creating a buffer zone of approach with soft vegetation and borders that acted as a kind of dentist's anaesthetic for those entering the park. Unlike the day-tripper at the seaside, the theme park visitor would hopefully banish all traces of normal life whilst experiencing their day out.

The concept of place was not removed without adding a substitute, and here Disney drew on his privileged position as a pioneer in the emerging mediums of film and animation coupled with the explosive take up of television to capture the cultural zeitgeist and emergent consumerist patterns of the time. History and nature could be reconstructed in a sealed environment and mindset, the power of media harnessed as the Disney theme park presented a themed view of life based upon the world of film and television. The absence of place is countered with a mediated state, linked to an 'external' world of film, television and viral branding. The returning visitor did not return to a place but to a heightened state of mind that 'made real' the ubiquitous and proliferating worlds of film, television and advertising.

This aspect grew to be the basis of successful franchises, culminating in spectacles such as Universal Studios who were branding theme parks in the 1990s by virtue of the fact that 'what is significant is less the geography of the park than its referencing of major motion pictures, which arguably have become more cognitively etched in the minds of visitors than geographical locations.[34] The Disney theme offered a 'double abstraction' – the theme park building a replica of an event, interpretation or environment drawn from a pre-existing 'tele-visual reality' (for instance, the Jaws-themed ride) – was based upon a recreation of the 'place' Amity as opposed to the real setting of the shark attack fictionalisation, Martha's Vineyard. This new world view merging the themed experience with the tele-visual reality soon came full circle, with films such as the *Pirates of the Caribbean* franchise stemming from Disney themed attractions.

## Conclusion

Battersea and Alton Towers can be considered as important and linked axis points in an understanding of British amusement spaces. Firstly, Battersea can be understood as a proto theme park, fermenting many of the elements that are considered as instrumental in the first Disney venture of 1955. That the singular theme – as experienced by the visitors – of Battersea and the Festival was Britishness is clear from its history. In turn, Alton Towers can be considered as an early example (certainly the first example in the UK) of the post theme park model – in effect, a proto post theme park. This model saw the development of parks utilising the megalithic structures of amusement technology (principally rollercoasters) associated with theme parks. In addition, Britishness permeates Alton Towers in its early literature, uniting it with Battersea. The Britishness of the 1980s Alton Towers experience sets itself against the United States, and this in turn links it back to internal anti-American rumblings associated with Battersea, creating a shared sense of inertia.

The recipe for a successful (post) theme park in the UK would seem to be evasive – essentially a copy of the American placelessness and high-octane thrill gathering, to give people both the thrill in of itself and the feeling of being in the correct environment to be thrilled. Eventually, possibly inevitably, the simulacrum's reference point vanishes and the themed experience desired is to simply be in a theme park. Whilst the modus operandi of a theme park, in terms experience, is to eliminate traces of the outside world, the theme park's engagement with the cultural forces in achieving this aim transplants the fantasy world within the park into spheres of everyday life. Like its contemporary analogy the virus, the theme park as a cultural context – a set of associations with regard to particular behaviour heavily branded and franchised mechanisms as part of general life – begins to spread. Alton Towers markets itself as the place to be from a vantage point of everybody wanting to be there.

## Notes

1  See, for example, the inside of the dust jacket promoting Barry Turner, *Beacon for Change: How the 1951 Festival of Britain Shaped the Modern Age* (London: Aurum Press, 2011).
2  Scott A. Lukas, *Theme Park* (London: Reaktion, 2008), p. 9.
3  For the chronological approach see Judith A. Adams, *The American Amusement Park Industry: A History of Technology and Thrills* (Boston: Twayne Publishers, 1991) and Raymond M. Weinstein, 'Disneyland and Coney Island: Reflections on the Evolution

of the Modern Amusement Park', *Journal of Popular Culture* 26, 1 (1992), pp. 131–64. Weinstein's ambitious work attempts to build a family tree between Coney Island, the first Disney park and the Disney EPCOT project, trying to establish ruptures and transitions. For a good example of a teleological approach see Tony Bennett, 'Hegemony, Ideology, Pleasure: Blackpool', in Tony Bennett, Colin Mercer and Janet Woollacott (eds), *Popular Culture and Social Relations* (Buckingham: Open University Press, 1986), pp. 135–54, which adopts Gramscian hegemony theory almost as a branded 'theme park' of radical theory in of itself.

4   An interesting approach to studying this tradition is given by Tony Bennett, *The Birth of the Museum: History, Theory, Politics* (London: Routledge, 1995) who extends the linkage around amusement spaces onto the museum as a part of a Foucauldian process by taking the displays of chaotic wonderment in the cabinet of curiosities and seeing them transformed into the regimented and assertive evolutionary spaces of modern knowledge. A more essentialist analysis is pursued by Steve Nelson, 'Walt Disney's EPCOT and the World's Fair Performance Tradition', *The Drama Review* 30, 4 (1986), pp. 106–46, who looks at the possible themes and space of the Chicago Columbian Exhibition of 1893 and draws comparison to Disney's futuristic EPCOT project.

5   Turner, *Beacon for Change*, p. 17.

6   Key references for the Festival are Becky E. Conekin, *'The Autobiography of a Nation': The 1951 Festival of Britain* (Manchester: Manchester University Press, 2003) and Mary Banham and Bevis Hillier (eds), *A Tonic to the Nation: The Festival of Britain 1951* (London: Thames and Hudson, 1976).

7   Mariel Grant, 'Working for the Yankee Dollar', *Journal of British Studies* 45 (2006), pp. 581–601 gives a detailed account of the delicate post-war relations between America and Britain and how the Festival tried to navigate this.

8   Turner, *Beacon for Change*, p. 87.

9   Philip Bradley, 'Festival Pleasure Gardens', *Fairground Mercury* 14, 3 (1991), pp. 7–14 details how the lease was negotiated as two years and then extended to five years. To facilitate this, the outlying Festival Gardens relinquished their admission fee and the fun fair – reduced in size from 9 acres to around 6.5 acres – retained a gate fee.

10   *World's Fair* (10 March 1951) and subsequent regular columns describing the fairground aspects of the Festival.

11   Naomi Games, *A Symbol for the Games: Abram Games and the Festival of Britain* (London: Capital History Publishing, 2011) details the story of the iconography and graphic identity of the Festival.

12   Lukas, *Theme Park*, p. 11.

13   Gesamtkunstwerk is a work of art that makes use of all or many art forms or strives to do so.

14   Conekin, *The Autobiography of a Nation*, p. 214.

15   A sense of irony haunts Priestley's words even further. If Priestley's *English Journey* is the start point for fellow socialist George Orwell's analysis of class and society *Road to Wigan Pier*, then recent events have seen the 'Pier' – an ironic (non) point of reference in Orwell's masterwork – re-imagined as a themed concept with branded retail areas.

16   Paul Rennie, *Festival of Britain Design 1951* (London: Antique Collectors Club, 2007), p. 70.

17   See for example: http://www.towersalmanac.com/history/index.php?id=2 accessed 8 July 2014.

18   See Gary Kelsall's project website on the heritage of Alton Towers: http://www.alton-towers.com/alton-towers-heritage accessed 8 July 2014; also http://list.historicengland.org.uk/resultsingle.aspx?uid=1000191.

19   *World's Fair* (8 September 1979), p. 6.

20   *World's Fair* (31 March 1980), p. 1.

21   *World's Fair* (24 April 1982), p. 4.

22   *World's Fair* (27 August 1983), pp. 6–7.

23   On the theme of the relationship between space and place see key works by Doreen
     Massey, *For Space* (London: Sage, 2005) and David Featherstone and Joe Painter (eds),
     *Spatial Politics: Essays for Doreen Massey* (Chichester: Wiley–Blackwell, 2013).

24   Foucault's development of heterotopias falls in and around his canon of key investiga-
     tions on knowledge, discipline, government of bodies, etc. There is not a key work
     that draws together his thoughts on this interesting topic but for a potential overview
     see Michel Foucault, 'Des Espaces Autres', *Architecture, Mouvement, Continuité* 5
     (October 1984), pp. 46–9: translation at http://foucault.info/doc/documents/heteroto-
     pia/foucault-heterotopia-en-html accessed 10 July 2014.

25   The author's PhD research includes work looking at the spatial aspects of these two
     types of layout. An enclosed fairground is akin to a labyrinth imprinted on a Mobius
     strip and achieves effect through visual aspects of repeating artwork inviting a topo-
     logical analysis; a street fair re-arranges open spaces, utilises interstices and creates
     new interstices inviting a topographical analysis.

26   See Tony Harcup, 'Re-imaging a Post-industrial City: The Leeds St Valentine's Fair as
     Civic Spectacle', *City* 4, 2 (2000), pp. 215–31 and Peter Stallybrass and Allon White,
     *The Politics and Poetics of Transgression* (London: Methuen, 1986). Bennett, in *Birth
     of the Museum*, tries to combine Foucault's idea of the heterotopia (a knowledge of
     space) with Foucault's earlier work on knowledge (a space of knowledge) to incorpo-
     rate the fluid space and radical potential of the travelling fairground into his theory of
     a general hegemonic direction towards fixed spectacle which moves from fairground,
     to amusement park, to museum.

27   Gary S. Cross and John K. Walton, *The Playful Crowd: Pleasure Crowds in the
     Twentieth Century* (New York: Columbia University Press, 2005) is currently one of
     the best sources that looks at the motivations of and counter-strategies against the sea-
     side crowd through parallel histories of Coney Island and Blackpool, feeding into the
     birth of the Disney phenomenon.

28   Josephine Kane, *The Architecture of Pleasure: British Amusement Parks 1900–1939*
     (Farnham: Ashgate, 2013), esp. ch. 3, which proposes the amusement park as a tran-
     sient space based upon the fluidity of the technological devices on offer. See also
     Josephine Kane, this volume. Alternatively it could be argued that the amusement park,
     much like a supermarket that changes its range of products as a matter of course, is a
     fixed space.

29   Bennett, 'Hegemony, Ideology, Pleasure' and Bennett, *Birth of the Museum* are key
     here, with a counter argument presented in Darren Webb, 'Bakhtin at the Seaside:
     Utopia, Modernity and the Carnivalesque', *Theory, Culture & Society* 22, 3 (2005),
     pp. 121–38.

30   Rob Shields, *Places on the Margin: Alternative Geographies of Modernity* (London:
     Routledge, 1991) with particular emphasis on his work on Brighton. A vivid science
     fictional rendering of the liminality of the beach can be found in J. G. Ballard's short
     story, 'The Reptile Enclosure' in *The Terminal Beach* (1964).

31   Bennett, *Birth of the Museum*, p. 210.

32   Adams, *The American Amusement Park Industry*, p. 87.

33   George Ritzer, *The Globalization of Nothing* (Thousand Oaks, California: Pine Forge
     Press, 2003) and Edward Relph, *Place and Placelessness* (London: Pion, 1976) cover
     this aspect amongst the many detractors of the Disney phenomenon. Marc Augé, *Non-
     places: Introduction to an Anthropology of Supermodernity* (London: Verso, 2009)
     includes wider examples of the onset of placelessness.

34   Lukas, *Theme Park*, p. 89.

# Part II
# International case histories

# 4  The heritage of public space

## Bondi Beach, Luna Park and the politics of amusement in Sydney

*Caroline Ford*

On a Friday evening in early November 1906, the Victorian Premier, Thomas Bent, opened Dreamland at St Kilda Beach.[1] Just six kilometres from the centre of Melbourne on the shores of Port Phillip Bay, and on the south-eastern corner of Australia, St Kilda was Melbourne's most accessible seaside resort. Since the 1850s, there had been a direct railway link to the city, and the town boasted a pier and sea bathing facilities. Although initially home to wealthy Melburnians, by the turn of the century trams brought city workers and their families directly to the beach. The local demography was also changing, as the grand houses left vacant in the wake of the 1890s Depression were transformed into more affordable housing.

To boost St Kilda's popularity, the local and state governments created a new leisure precinct on public land. Dreamland lay at the centre of this new destination. Visitors to Dreamland rode on an airship to the moon, and on boats through the major rivers of the world. Thrill-seekers could also slide from the peak of the 30 metre-high Mount Fujiyama, experience Heaven and Hell in Hereafter and, just six months after the real thing, a version of the San Francisco earthquake. Like the American amusement parks on which it was based, Dreamland was decorated with 'myriads of gleaming tiny electric lights' and beamed searchlights into the sky at night.[2] It brought tried and tested American novelties to Australian shores, many for the first time. But it was not the success many hoped or expected.

One month later, more than seven hundred kilometres north on the country's east coast, the New South Wales (NSW) Premier, Joseph Carruthers, opened Sydney's own beachfront amusement park. Wonderland City at Tamarama Beach was the brainchild of William Anderson, a theatre manager from Melbourne. It boasted similar attractions to those found at Dreamland including Rivers of the World boat journeys, airships and side shows.

Many in Sydney already associated Tamarama with seaside entertainment. Seven kilometres east of the city and next to the more famous and larger Bondi Beach, Tamarama was a surf beach in contrast to Melbourne's more gentle bay beaches. Two decades earlier entrepreneurs had constructed a British-style aquarium and pleasure grounds on this site, mixing amusements and sideshows with more educative forms of entertainment at a spectacular coastal location. Although popularity waned in the 1890s, the aquarium had lasted until the early 1900s.

It was from the ruins of this enterprise that Wonderland City now rose. Unlike Dreamland, Wonderland City was on private land and was not a state government initiative, although it initially had government support.[3]

Dreamland and Wonderland City were not just inspired by American seaside amusement parks, they were replicas: in name, design, promotion and the particular attractions they offered. They were local attempts to exploit the popularity of the new amusement parks emerging around the American coast, most notably at New York's Coney Island, and were part of a worldwide shift in commercial amusements. Although Wonderland City partly resembled the site's earlier aquarium and pleasure grounds, the aquarium had been inspired by British rather than American models of seaside entertainment. Australia would soon develop its own beach culture, which would come to influence seaside cultures across the world, but for now, it was importing beachside amusements.

Yet, while both amusement parks initially attracted strong crowds they failed to maintain enough interest to keep them viable over the long term.[4] Dreamland lasted just one season, and by February 1908 many of the buildings had been sold. The site remained enclosed by an 'unsightly' black fence, derided by the Premier and residents as an eyesore, a 'desolate, dreary, forlorn looking piece of ground' that would be 'a disgrace to any municipality', although the Figure Eight railway was retained and continued to be popular.[5] Wonderland City lasted slightly longer. Through constant reinvention and well-publicised stunts Anderson kept Wonderland City afloat – just barely – for five brief summer seasons. By 1911, the park had permanently closed. Big storms the following year destroyed much of the remaining infrastructure, and a decade later residents won the public park behind Tamarama Beach they had long fought for.[6]

Dreamland and Wonderland City were not unique in their short lives. Amusements are by their nature transient, requiring ongoing maintenance and failing over the long term to satiate crowds always looking for new thrills.[7] But in Sydney, where the weather is more appropriate for swimming for a longer portion of the year, there was an added obstacle: these amusements were emerging at the very time surf bathing and sun bathing were becoming mainstream activities. Sydney's beachgoers now sought their thrills not on rides but in the surf and on the sand, for free.

This was especially a challenge for Wonderland City because it was partly built on a beach that was a Crown reserve. Earlier state governments had rewarded the former aquarium's commercial enterprise by approving exclusive use of this beach reserve but now Anderson encountered resistance from both state and local governments. He inadvertently became the centre of a long-running battle over free public access to local beaches just as political favour for private interests was turning. Ironically, Wonderland City helped to consolidate opposition towards commercial amusements just as politicians began to agree the city's beaches should be free public spaces. Here, the debate culminated in the state government ordering Anderson to remove the fences that allowed him to charge for access to Tamarama Beach in 1907. In reluctantly complying, he lost one of Wonderland City's most valuable assets.[8]

The local opposition to Wonderland City was part of broader debates in both Sydney and Melbourne about whether the commercialisation of beaches was appropriate, how the management of beach parks should be funded and what the general public expected a beach to be. The challenges and concerns were similar in both places but the government and local responses were starkly different. In Melbourne, the Victorian state government sought to enhance the commercial and entertainment opportunities of the St Kilda foreshore area while in Sydney the lesson from Wonderland City was that the NSW government was firming in its opposition to coastal amusement parks. Soon, this principle was applied to almost any alienation of coastal parks for commercial purposes. The backlash against Wonderland City's occupation of Tamarama Beach had sparked a debate that consolidated a particular way of thinking about commercial ventures on beach-front land that would have lasting policy implications.

Yet commercial amusements continued to appeal to local councils as they struggled to manage beach parks and provide services for the growing numbers of residents and tourists who wanted to enjoy those beaches. As more coastal areas were dedicated to public recreation and became Crown land managed in trust by local councils over the early twentieth century, the need for consensus between state and local authorities became critical. But as this chapter shows, Sydney's beachfront parks came to be seen as sanctified and inviolable, in many cases preventing the creation of new beachfront amusement parks, despite ongoing commercial and local government demand for private lease opportunities.

## Shaping the beach

After Dreamland closed, its Figure Eight railway remained popular among visitors to St Kilda but the rest of the site drew criticism, including from the government. The St Kilda Foreshore Committee, a group of local and state government representatives that had been formed to improve the St Kilda area, favoured continuing commercial amusements on the site. It saw benefits both in the income attained through amusement leases and in the likelihood of attracting more tourists to the bay-front suburb. As part of larger works around the St Kilda foreshores, it leased the Dreamland site to American showmen who built, in 1912, a Luna Park. Like Dreamland before it, St Kilda's Luna Park closely imitated the Coney Island model. Its owners, J. D. Williams and brothers Herman, Harold and Leon Phillips, selected the most popular rides and amusements then in the United States, and imported a team of twenty American funfair designers to create the park. Extravagantly adorned with 80,000 electric lights, and with the entrance beneath a huge neon-lit 'Mr Moon' face, the park provided thrill rides and live entertainment (Figure 4.1). It would reinvent itself several times over the following decades to rejuvenate flagging fortunes, increase appeal to the Melbourne population and ensure longevity. More than a century later, the park's iconic moon face continues to greet visitors as they step off the tram and enter Australia's longest running amusement park. It is a part of St Kilda.

*Figure 4.1* The iconic 'Mr Moon' face has welcomed visitors to St Kilda's Luna Park since it first opened in 1912

Source: © State Library of Victoria

While the St Kilda Foreshore Committee was actively seeking commercial investors and trying to build St Kilda's appeal as an entertainment district on various international models, consecutive NSW governments were firming in their opposition to commercial amusement leases on Sydney's beach front parks. Wonderland City had been the first victim of a shared public and government anxiety over commercial occupation of public coastal spaces. Others soon followed.[9] A 1913 application by Williams and the Phillips brothers to replicate St Kilda's Luna Park on Sydney's Bondi Beach confirmed the NSW state government was still opposed to the principle of commercial amusements on the beaches. Just two years earlier this Labor government had invested in a high-profile foreshore resumption scheme to claim back harbour and beach foreshores to facilitate 'healthy' recreation for the city's workers. It was now reluctant to approve a venture that might compromise free public access to a beachfront park. Despite acknowledging that Bondi Park was so large the lease was unlikely to raise serious local objections, that the amusement park would improve a portion of the park that had received little attention, and that the local Waverley Council would benefit considerably from the lease revenue, staff in the Department of Lands described the proposed lease as 'utterly opposed to the object for which these areas are provided'.[10]

The lease was not granted on a Crown Solicitor's advice that it was illegal. But it was apparent that in NSW, commercial amusement parks were seen as a threat to rather than enhancement of public values of open areas. Similar rulings against amusement and sporting leases along the beachfront and in other public parks in the following years confirmed this government would not interfere with open public spaces.[11] Their pro-park philosophy was drawn from the nineteenth-century rationale, originating in Britain, which promoted public parks as primarily open spaces for healthy recreation for urban workers.[12] Spending money on cheap thrills was not deemed a 'healthy' activity, nor were amusement parks considered a 'healthy' landscape.

## Fleeting amusements

The Luna Park did not eventuate at Bondi but an amusement park was soon built near the Harbour at Rushcutters Bay, closer to the city. Built on the site of a Chinese market garden in 1913, White City was the enterprise of Australian rather than American entrepreneurs; although like St Kilda's Luna Park it was based closely on overseas models. Likely named and modelled after the famous Chicago amusement park, its owners employed T. H. Eslick, who had worked on Melbourne's Luna Park and others around the world, to construct the park.[13] Despite their efforts to be true to American models, and the amusement park's apparent popularity (as reported in the press), White City lasted about as long as Wonderland City had. A severe storm in September 1917 completely destroyed one of the entrance domes and caused extensive damage throughout the park.[14] It was never rebuilt, and the park went into liquidation in 1918. In the early 1920s, the White City site was transformed into a tennis club bearing the same name.

Sydney's brief but enthusiastic dalliance with large American-style amusement parks appeared to be over.

Closer to the beaches, small-scale amusements proliferated, providing night-time entertainment for Sydneysiders who sought a particular brand of escape at the beach. Waverley Council had the authority to approve small events in Bondi Park but in the early 1920s, unable to police the growing numbers of fairs and carnivals held in the park by voluntary organisations, it stopped granting permission.[15] However, amusements continued to operate, often unregulated, on private land around the fringes of the park at Bondi, Manly and other popular beaches.

In the 1920s, local entrepreneurs, inspired by the global resurgence in amusement park popularity, again tried to build amusement parks on the edges of the city's most popular beaches but most were blocked by government resistance or financial struggles. Towards the end of the decade, however, their persistence was rewarded, and it increasingly appeared as though commercial enterprises, and particularly amusement parks, were the future of Sydney's beaches. The Randwick Council set the bar, approving a British-style pleasure pier to be constructed from the centre of Coogee Beach. Marred by controversy and poor management from the outset, it never secured a place in Sydney's popular leisure scene and was demolished in the early 1930s. But in the late 1920s, the pier, and more significantly the shark enclosure the Council strung from it in the wake of several fatal shark attacks, boosted Coogee's popularity and the Randwick Council's income, providing a strong motivation for the other councils to develop their own local amusement economy. At Manly, a thin peninsula with a large surf beach on one side and more gentle harbour beaches on the other, a harbour-side pier was transformed into the 'Manly Fun Pier', Sydney's own version of the British pleasure piers, albeit on a tiny scale. Lasting into the 1980s, it would be much more successful than the more extravagant Coogee version.[16]

Among all the attempts to construct major amusement parks in Sydney in the inter-war period, three stand out as particularly significant moments in the heritage of amusements in Sydney: at Bondi Beach (Waverley Council), Maroubra Beach (Randwick Council) and Milsons Point (North Sydney Council). Only the latter was successful and, significantly, it was on the edge of Sydney Harbour, kilometres inland from the city's famous ocean beaches. Nonetheless, the failed attempts to introduce major amusements like the St Kilda Luna Park at Bondi and Maroubra beaches, and the attempts to stop the construction of the Luna Park at Milsons Point, reveal important insights into the complex meanings attached to both amusement parks and public spaces in Sydney in the early 1930s.

## Fighting for a Luna Park for Sydney

The battle for a Luna Park at Bondi took place over four years from 1929. In the second half of the 1920s, as the Coogee pier and shark net were materialising further south, the local Waverley Council invested £160,000 of its own funds into its Bondi Improvement Scheme, a series of works designed to improve the residential and tourist appeal of the beach suburb that included redesigning Bondi

Park and building the now famous Bondi Pavilion. It had sought state government assistance, arguing that people from outside of the local borough would benefit the most, but the state had insisted this was a local issue. As the works were nearing completion, the Council began to consider an amusement park for the largely deserted and unimproved southern corner of Bondi Park. It would both provide a useful income stream for the Council through a long-term lease, and rejuvenate this part of Bondi at no cost to the Council. It would transform night-time Bondi, according to one aspiring lessee, 'from a morgue into a place of amusement'.[17]

The Council had received a petition with 1,100 signatures supporting an amusement park but it underestimated local opposition. The Council itself was also divided, with dissenting aldermen denouncing the proposal for commercialising part of a public park. The state government shared their concerns, and as the park was state Crown land managed in trust by the Council, the lease required Ministerial approval. After three years of intermittent internal debate, by 1932 the Council was becoming increasingly desperate as it counted the financial cost of lower than anticipated use of its extravagant pavilion. But on the recommendation of the Metropolitan Land Board, which had held an inquiry into whether there were 'any objections in the public interests' to a long-term amusement lease on Bondi Park, in 1933 Ernest Buttenshaw, the conservative Minister for Lands, announced he would not approve an amusement lease at Bondi Park. Michael Bruxner, the acting Premier, endorsed his decision, insisting there would be no 'further alienation of the people's preserve' at Bondi Beach.[18] The preference for open spaces over commercial amusements, now several decades old, continued to prevail at Bondi.

Yet elsewhere, the government was more willing to consider the merits of an amusement park on public land. At Maroubra Beach, around six kilometres south of Bondi and little more than ten kilometres from the city, it was even willing to compromise free access to a public park to secure a commercial lease. In 1935, Buttenshaw approved a 28-year amusement lease for a park at Maroubra Beach that would 'rival Coney Island', again on the recommendation of the local land board. In complete contrast to Bondi, he supported the proposal despite the local Randwick Council's opposition, suggesting that for the state government, the potential commercial benefit and boost to local development outweighed any environmental or social concerns. However, the amusement park was never built, likely due to financial obstacles for the lessees.[19]

When Sydney did get its own major amusement park in 1935, it was built not on the beach but on the city's harbour foreshore, directly across the water from the city's busiest ferry terminal. Behind the north pylon of the then three-year-old Sydney Harbour Bridge at Milsons Point, the syndicate behind St Kilda's Luna Park finally opened their long-desired Sydney franchise with full government support after more attempts to invest £50,000 in a Luna Park at Coogee in 1925[20] and £75,000 at Bondi in 1932.[21] They had also opened a Luna Park at Adelaide's Glenelg Beach in 1930 but that had by now failed, a victim of lower-than anticipated profits (partly due to the Depression) and resistance from the local council (Figure 4.2). Sydney's famous Big Dipper rollercoaster came from the Glenelg Luna Park, dismantled and shipped to Sydney Harbour where it was reassembled

*Figure 4.2* Luna Park formed a brief but controversial backdrop to Adelaide's popular Glenelg Beach in the early 1930s

on site. Herman Phillips, the lessee of Sydney's new Luna Park and David Atkins, its manager, were Australia's amusement elite: Atkins had also been associated with the amusements at Manly Pier, Coogee Ocean Pier and the Glenelg Luna Park. And they guaranteed Sydney's Luna Park would be world class.[22]

When the first visitors clicked through the turnstiles beneath the 'gargantuan teeth' of Sydney's very own laughing face in October 1935, the city reportedly 'put on its old clothes and became young again'. This was not a unique amusement park; Phillips and Atkins were determined to build a park containing some of the most popular shows and rides then in use in America, so, like those before it, Sydney's Luna Park closely emulated the Melbourne and international models. It boasted amusements that could be found in other contemporary amusement parks throughout America, Britain and elsewhere in Europe including dodgem cars, a ghost train, Noah's ark, river boats, slides and a 'goofy house' (Figure 4.3). As had long been a tradition with amusement parks here and elsewhere, at night the buildings and grounds were 'brilliantly illuminated'.[23] But nearly twenty years after White City disappeared from the city's tourist brochures, this was Sydney's own amusement park. The night-time view of the park's sparkling entrance towers and amusement buildings reflected on the harbour waters soon became a defining image of Sydney's Luna Park, and of the harbour's nightscape when viewed from the city.

## Resisting Luna Park

The proposals for all three amusement parks divided the local communities. Unsurprisingly, they all attracted support from local investors, hopeful lessees and people who would use the parks. At Bondi, the local *Eastern Free Press* argued an amusement park would 'brighten up' a beach that resembled 'the sands of darkness' at night.[24] At Milsons Point, the local and state governments welcomed the Luna Park for its promise to rejuvenate an area that had been abandoned during bridge construction, and to liven up what had been an industrial site.

Yet, the proposals also attracted strong criticism from particular sections of the community who mobilised against the amusement park. Debates were played out in the press, at council and public meetings, through petitions and at the Land Board Inquiries at Bondi and Maroubra. Opponents cited the failures of White City and Wonderland City decades earlier as evidence that amusement parks were not suitable for Sydney. At Bondi, local residents and at least one alderman formed the Bondi Beach and Foreshores Defence Committee to co-ordinate their campaign against the park, and immediately registered the group's 'emphatic protest' against the proposed lease. The Bondi State Electoral Conference, North Bondi Progress Association and Parks and Playgrounds Movement of NSW passed similar resolutions and wrote to the Minister for Lands, all concerned about the potential loss of public park lands. At Maroubra, the Randwick Council even opposed the lease it had initially supported (although it remained divided), concerned it might have a negative local impact.[25]

*Figure 4.3* The Spider at Sydney's Luna Park, 1938. The Luna Park and Harbour Bridge are intrinsically connected: built within three years of each other, the two icons frame the view across the harbour from the city

Source: © State Library of New South Wales

At both Bondi and Milsons Point, a group of influential Sydney architects, artists, historians and nature-lovers had mounted persuasive arguments against allowing the amusement leases. Through groups including the Town Planning Association, Tree Lovers' Civic League and Parks and Playground Association these individuals worked together in the inter-war period to conserve Sydney's natural and historic heritage, including parklands, in the face of rapid suburban development.[26] Although the proposed Luna Parks were contained in far smaller parcels than the land they sought to protect in their more prominent campaigns, which included national parks north and west of Sydney, the proposed amusement parks were among many issues these individuals and groups feared would threaten their values of public spaces at the local level.

In all the complaints against the proposed amusement parks and state and local government concerns, a number of themes emerge. Firstly, there were social concerns: amusement parks were associated with working-class forms of entertainment that many local residents and businesses did not want in their backyards. For this reason amusement parks also attracted the ire of middle-class reformers who sought more 'healthy' forms of recreation for the city's workers and their families. The Randwick Council was concerned that an amusement park at Maroubra might attract 'undesirable persons' just as the *Truth* argued amusements would bring the 'lowest type of human vermin' to Bondi Beach.[27] A NSW surveyor sent to Melbourne to research the St Kilda Luna Park to inform the government's decision at Bondi returned with positive reports about the 'orderly behaviour' of the crowd there in a specific response to such negative perceptions.[28] At Glenelg, criticisms that the Luna Park was a 'haven for carnival types and undesirables' were endorsed by the local council and had contributed to its downfall.[29]

Local clergy were particularly vocal in protesting the proposed amusement parks at both Bondi and Milsons Point on this basis. Six local ministers condemned the Bondi proposal for its potential to attract an 'undesirable class' of people to the part of the beach popular with families and children, while Revd Calder from North Sydney described Luna Park as 'a menace to the morals and well-being of the people of the district' that would likely become the site of 'nightly orgies'.[30]

Concerns about the 'type' of people they might bring to the beaches had been a common theme in arguments against amusement parks and amusements over the previous two decades. In an era when politicians, eugenicists and bathers themselves promoted beach recreation as healthy for the mind and body, cheap amusements were seen to appeal to a more troublesome group of working-class thrill-seekers. The centuries-old British stigma that condemned recreation of the 'masses' as unrestrained, unproductive and dangerous to healthy minds and bodies, although by now fading, continued to influence amusement detractors in Sydney as elsewhere.[31]

In this context, opposition to the construction of an amusement park on the shores of one of the city's most popular beaches can be viewed as at least partly driven by a patriarchal concern for the welfare of working-class leisure seekers if not by an inherent criticism of those 'types' of people. An amusement park might have corrupted an otherwise innocent and healthy day at the beach. Given intellectuals in America were particularly critical of cheap amusements, it is likely the Sydney architects, town planners and park advocates who rallied against local amusement parks shared a disdain for this form of recreation – although they did not specifically articulate one. But many coastal residents also did not want Sydney's working-class thrill-seekers descending on their neighbourhoods. There was also undoubtedly a class element to the government's preference for amusements at Milsons Point and Maroubra over Bondi, for the former two suburbs were home to working-class residents and did not attract middle-class leisure seekers on the scale of Bondi.

A second concern focused on the aesthetic presence of amusement parks, criticising their appearance and the loud noises they would generate. The prospect of rides and structures lit up at night was particularly galling to some. The Randwick Alderman Thomas Mutch expressed the views of many when he described the Luna Parks at St Kilda and Glenelg as 'hideous'.[32] Amusement park critics including Randwick and Waverley aldermen and, in the case of Bondi, the state government also argued the appearance and noises of amusement parks would harm rather than boost local economies and the appeal of beachfront living.[33]

Aesthetic issues were particularly prominent in the debates about the proposed Bondi amusement park. In the minds of many, Bondi Beach's visual appeal had been enhanced by the clean modern lines of the new sea-wall, promenade, marine drive and park recently constructed by the Council. It was the appearance of this landscape, as well as the landscape itself, that many fought to protect. John Sulman, an esteemed architect and former president of the Town Planning Association (TPA), who had helped select the new Bondi Pavilion design close to a decade earlier, warned that the proposal would spoil the coastal view and 'be a detriment to the whole of Bondi Beach, which we all recognise is one of the finest on the NSW coast'. Leslie Wilkinson, Professor of Architecture at the University of Sydney, agreed that an amusement park, no matter how 'architecturally beautiful', would take away from the area's 'natural beauty', while fellow architect and secretary of the TPA, Bertram Ford, deemed it 'vitally important' that no amusements blocked the view of the ocean from the road. William Bailey, Chairman of the National Park Trust, described the proposed amusement park as a 'blot on the landscape'.[34] These statements and others like them point to the existence and influence of a group of professionals supportive of natural landscapes – or of landscapes carefully crafted to 'enhance' the natural. But their perspectives were shared by many regular beachgoers and city residents.

Thirdly, environmental concerns were also aired, particularly at Bondi, where the Metropolitan Land Board Inquiry had investigated any potential impact an amusement park might have on trees. The area earmarked for the amusement park contained 52 Norfolk Island pines, like those that famously stretched across the Manly ocean front. These were not large trees but they provided shelter for park users in summer. They varied in height from 10 to 35 feet and were around 22 years old, having been planted in an earlier phase of municipal investment in beautifying Bondi Beach. Charles Bean, Secretary of the NSW Parks and

Playgrounds Movement, and an esteemed former national war historian, had first alerted the Minister for Lands to the threat a potential amusement park posed to 'trees planted twenty years ago for beauty and to give shade for the people', and they became a major theme of the inquiry.

Annie Wyatt, a member of the Forest League and co-founder and secretary of the Tree Lovers' Civic League, was particularly outspoken on this issue. A nature-lover who had lectured on tree cultivation and spoken about forest conservation on radio, Wyatt was concerned about the impact of crowds and construction on the tree roots, and the potential for buildings to block sunlight from the trees. Others echoed her concerns.[35] The aspiring amusement park lessees who spoke at the Inquiry were all careful to reassure the Board their proposals would not damage or threaten the park's trees. Yet, in recommending against the lease one of the Land Board's key findings was that a park 'must assuredly have a bad effect on the trees which at present render the area attractive to the users and those visiting Bondi'.[36]

## Fighting commercialism

By far the most prominent and (at Bondi) most effective objection against the proposed amusement parks emphasised their commercial nature at the expense of free and open public parks. By the early 1930s, Sydney had a strong record of more than two decades of bipartisan political support for non-commercialised parks, particularly at the beaches. What had begun at Tamarama in 1907, with a government decision to force Wonderland City to provide free access to the beach, had soon become a shared principle underpinning attitudes to the city's foreshores – that they be freely accessible to all. The idea that beaches were critical in providing clean and healthy recreation for the city's workers had come to underpin political attitudes towards the coast, making protection of existing reserves even more important. The only commercial enterprises that were widely supported were those that facilitated surf-based recreation, such as dressing sheds and beach hire businesses.[37]

At Bondi Beach, the potential commercialisation of a corner of the public reserve was the government's primary concern. A Department of Lands report into the proposal condemned amusement parks for replacing 'natural conditions and open spaces' with 'artificial and garish amusements'. The Metropolitan Land Board found 'serious objections in the public interest' in the Bondi proposal, which included that 'the reserves in the vicinity of the beaches are by no means excessive and any curtailment should be opposed, the construction of amusements would assuredly affect the free use of the reserve' (Figure 4.4). John Sulman also branded the proposal a 'crime' for this reason. Minister Buttenshaw's decision not to permit the Bondi amusement lease was driven by the government's determination to protect against any 'further alienation of the people's preserve'.[38]

Yet, at Maroubra, in the face of the same defences, Buttenshaw and the Metropolitan Land Board were both willing to sacrifice existing public park space to commercial amusements. The government's commitment to protecting public

*Figure 4.4* At Bondi Park, behind Australia's most famous beach, local residents fought the construction of an amusement park in favour of open parkland and shade-providing trees

Source: © National Library of Australia

parks was more flexible than it would admit. This decision, and the approval of the Coogee pier lease in 1923, which had been granted in the face of local opposition, were the exceptions to a trend of state government protection of ocean beach parks.[39] From the popular perspective, all beaches were equally worth protecting and preserving. But the anomaly was justified, in the state government's eyes, because Bondi and Maroubra beaches held different values.

Both beaches were surrounded by large public reserves, created in 1881 and 1909 respectively, to cater to imagined future crowds. But while by the early 1930s Bondi attracted tens of thousands of people on hot summer weekends, and was becoming an icon of the city and nation, Maroubra had been left behind.[40] Although both the beach and park were larger than at Bondi, Maroubra was further from the city, and had only been directly accessible by tram since 1921. The state government had invested in improving its own land holdings surrounding the public reserve in the 1920s, flattening the sandhills, filling in swamps and evicting local market gardeners in a bid to increase land values and subdivide and sell its vast holdings in the area. But the state and local governments and local investors were still awaiting the great residential boom that had occurred at other beach suburbs including Bondi. An agreement between the state and local governments in the late 1920s to sell part of the park in exchange for building a marine drive along the beachfront and other improvements to the beach had come to nothing, so there remained little, except the beach itself, to draw tourists and residents to Maroubra.[41] And Sydney had many beaches.

The state government's incentive for permitting an amusement park at Maroubra Beach was therefore far greater than any that may have existed at Bondi. The Waverley Council may have been desperate to boost Bondi's attractions but from the state government perspective it was popular enough; indeed, the government branded the Waverley Council greedy for its money-grab.[42] Maroubra Beach, however, needed a boost, and the government stood to benefit from growth in its residential and investment appeal. But Bondi also held special significance: it had a place in the popular imagination that experts and governments were willing to defend. Maroubra did not.

At Milsons Point, the same professionals who had mobilised at Bondi, again fought for the foreshore area earmarked for the Luna Park to be reserved as a public park. They joined representatives from the local progress association, local aldermen, parliamentarians and church representatives to present a petition signed by 1,200 residents to the Assistant Minister for Local Government demanding the public park.[43] John Bradfield, who had designed the Sydney Harbour Bridge, had also anticipated civic spaces on the harbour foreshores in the wake of its construction.[44] But this was not an existing park or a popular beach like those at Bondi or Maroubra beaches, and Luna Park's detractors were unable to persuade governments at any level to preserve the area for open access. It was an unused industrial area, the former site of workshops associated with the construction of the Sydney Harbour Bridge. The proposed amusement park did not threaten existing public use but rather limited that possibility for the future.

From the government and council's perspective, an amusement park, together with the planned construction of an Olympic swimming pool next door, would bring people back to the north shore of the harbour following considerable disruption during bridge construction, and perhaps encourage use of the bridge. They insisted ample open parkland already existed in the area, and any extension of that parkland would increase the financial burden on the Council. Responding to criticisms about favouring commercial amusements over public parks, the Premier Sir Bertram Stevens reassured parliament that his government would 'preserve the rights of the people and . . . act at all times in the public interest'. But here, as at Maroubra, public interest was defined more broadly than preserving existing public space. It was interpreted in favour of commercial imperatives.[45]

## Defending amusements

Following its opening, the local North Sydney Council received regular complaints about Luna Park: residents complained about the loud noise created by the park's amplified music, leading to stricter rules about noise after hours; local businesses complained about competition from street vendors; and religious groups demanded the park be closed on Sundays. A petition of more than 1,200 signatures supporting Sunday openings meant the latter were unsuccessful.[46] Over the following decades, the park's popularity wavered. In 1969, the park was sold to a company that applied to develop the site as a world trade centre with seven multistorey buildings along the foreshore designed by local architect Harry Seidler. But the state government did not approve the alternative use, which it deemed unsuitable for the harbour foreshore (Figure 4.5). The company continued to operate the Luna Park with minimal investment, leading to degraded features. Artist Martin Sharpe, who was employed to repaint some of the park features, later recalled it felt as though he and fellow restorers were 'decorating a rotting cake'.[47] But, while they may not have been visiting the park in large numbers, Sydneysiders had become attached to its presence, and to the appearance of its smiling face beaming across the harbour, brilliantly illuminated at night.

In 1979, following several safety incidents and the horrific deaths of six children and one adult in a fire in the ghost train, Sydney's Luna Park closed. The following year, the state government appointed a new lessee who auctioned many movable items and demolished and burnt the Big Dipper, Davey Jones' Locker and River Caves. A somewhat diminished park reopened in April 1982 and limped through the early 1980s, opening only on weekends and school and public holidays.[48] Over the following decade, the Council and state government battled over the site. Several developers keen to build residential towers approached the Council, earning the ire of residents who opposed high-rise buildings on the foreshore and the many who wanted Luna Park preserved. It appeared that Sydney's Luna Park was close on the heels of many other long-standing major amusement parks that had closed in the changing social and cultural context of the inter-war decades. What saved Sydney's park, ironically, was community action. And it was the city's artists, architects and conservationists who now fought for its survival.

*Figure 4.5* The Big Dipper at Sydney's Luna Park operated from 1935 to 1979. Thrill-
seekers sought amusement on the harbour foreshore, a site politicians deemed
far more appropriate than the sands of Bondi Beach

Source: © State Library of New South Wales

At a public meeting of the Friends of Luna Park in 1980, several high-profile
speakers advocated for the retention of Luna Park. Peter Johnson, Dean of the
University of Sydney's Faculty of Architecture, described the park, with its fantasy architecture and 'important series of artefacts', as 'an essential element which
contributes to the identity of Sydney'. The leading Sydney architect Harry Seidler,
who would later mount a legal challenge against the noise pollution created by the
park's rollercoaster, stated that he occupied an office overlooking Luna Park and
that 'over the years I have grown to love it dearly'. Echoing Annie Wyatt, he also
expressed concern for the trees that surrounded the park if development were to
occur on the site. Jack Mundy, who earned national fame during the 1970s as the
union leader of the 'Green Bans' on demolishing heritage buildings and maintaining green space, insisted on behalf of the Australian Conservation Foundation

that the government had a 'social responsibility to maintain the essential spirit of Luna Park'. Howard Tanner from the National Trust, the very organisation tree-lover Annie Wyatt had helped to create nearly half a century earlier, and which had listed the face and towers as having heritage value, also spoke in favour of retaining the park and recognising its architectural significance to the harbour. Leo Schofield, a journalist, advertiser and food critic attending the meeting at the impetus of his children, argued the park was 'integral and basically irreplaceable part of the Sydney harbourscape' and inherently Australian.

The meeting called for the site to continue to be used as an amusement park called Luna Park, with the façade and as much of the current equipment as possible to be retained. It also sought government assurances that the large trees skirting the park would be preserved.[49] Nine years later, a public meeting of more than 200 people passed a similar resolution. Mobilised by the government's revocation of council's planning powers over concerns about plans for inappropriate development, they demanded there be no private development on the Luna Park site, and that its character and essential features be retained and the land dedicated as a public park.

Remarkably, these architects, artists and conservationists were the modern equivalent of those who had fought against the Luna Parks during the 1930s. In some ways their arguments were similar: that the land and foreshore area should be publicly accessible, and important trees protected. They, too, were motivated by a fear of losing a space they deemed important to the life of the city. But they wanted to *preserve* Luna Park. Against the threat of high-rise buildings, the amusement park had become a public space worth protecting. But more than this, their and others' defence of the site betrayed an attachment bred from familiarity: Sydney's Luna Park had become an accidental monument of the city.[50] And ironically, given earlier concerns about the 'type' of people an amusement park might attract, the park's heritage value was linked to its working-class associations. This was vernacular heritage at its most fun.

Conceding to public pressure the government decided to retain the amusement functions of the site, legislating for it to be vested in the Crown and dedicated for public recreation, amusement and entertainment. The government also commissioned a heritage study that concluded that Luna Park was 'an outstanding item of environmental heritage significance'. It recommended Luna Park be retained in public ownership and conserved and adapted for continuing use as a public amusement park.[51] Sydney's Luna Park was reinvented and reborn. It resumed its place at the heart of the city, not as a private commercial space but as a cherished public place of amusement on the edge of the harbour.

## The heritage of amusements and public parks in Sydney

Bondi Beach and the Luna Park at Milsons Point are both listed on the NSW state heritage register. This means they have been identified as holding state-level heritage significance, and their heritage values are protected. Bondi Beach is recognised for its cultural landscape, which contains 'all of the elements which are typical of the Australian beach: a pavilion for public changing rooms, surf life-saving club(s), wide expanse of sand, grassy park for picnicking, a promenade/marine drive, and the ability of ocean swimming'.[52] Luna Park is recognised for its social, aesthetic and architectural significance, and as a 'landmark on Sydney Harbour'.

It is the landscapes that were primarily shaped by government decisions and commercial pressures in the 1930s that are celebrated and retained through these heritage listings, landscapes that primarily represent nature, at the beach, and culture, on the harbour. Whereas Luna Park is an urban icon, a seemingly intrinsic part of the view north from the business centre that includes the equally iconic Sydney Harbour Bridge, Sydney Opera House and Sydney Harbour, Bondi Beach is famous for its cultural landscape: the surf, sand and grass, framed by a concrete wall, paths, roads and a promenade. All of Sydney's ocean beaches are built around sun and surf. There have been many amusements, funfairs, rides and circuses, but there has never been an amusement park on the scale of Wonderland City or White City since the former was abandoned more than a century ago.[53] There is no room at Bondi, the 'typical Australian beach', for a permanent amusement park.[54]

Yet, more than just landscapes, the heritage listings also indirectly recognise and protect the cultural uses of these two places. Their heritage values reflect and are driven by strong local cultural attachment to both these places.[55] For Sydney residents today, the thought of a beach bordered by commercial amusements is as alien as the thought of the corner behind the northern pylon of the Harbour Bridge being home to anything other than a neon-lit smiling face. Locals have grown attached to both these spaces as they have known them to exist, and fought to protect their existing uses. Sydney's ocean beaches are a monument both to a particular type of leisure and to an ethos of public access. There is no place on the ocean beaches for an amusement park, but the amusement park, just like Bondi Beach, has taken its place in the heart of the city.

## Notes

1 This chapter is based on research supported by a New South Wales Archival Research Fellowship. With thanks to Catherine Bishop, Mark Dunn, Hannah Forsyth, Toby Martin, Lisa Murray, Alex Roberts and Richard White for their thoughtful advice on an earlier version of this work.
2 *Prahran Telegraph* (27 October 1906), p. 2.
3 Caroline Ford, 'A Summer Fling: The Rise and Fall of Aquariums and Fun Parks on Sydney's Ocean Coast 1885–1920', *Journal of Tourism History* 1, 2 (2009), pp. 95–112.
4 Sean Brawley, *The Bondi Lifesaver: A History of an Australian Icon* (Sydney: ABC Books, 2007), p. 25.
5 *Argus* (9 June 1908), p. 5; *Prahran Telegraph* (18 July 1908), p. 2.
6 Caroline Ford, *Sydney Beaches: A History* (Sydney: New South Books, 2014).
7 Gary S. Cross and John K. Walton, *The Playful Crowd: Pleasure Places in the Twentieth Century* (New York: Columbia University Press, 2005).
8 Ford, *Sydney Beaches*.

 9  See Ford, 'A Summer Fling'; Pauline Curby, *Seven Miles From Sydney: A History of Manly* (Sydney: Manly Council, 2001), pp. 180–81.
10  State Records NSW (hereafter cited as SRNSW): Department of Lands, Miscellaneous Branch; NRS 8258, *Letters received 1867–1979* [13/6813].
11  SRNSW: NRS8258 [14/14167].
12  Hazel Conway, *People's Parks: The Design and Development of Victorian Parks in Britain* (Cambridge: Cambridge University Press, 1991), pp. 2–3.
13  *The World's News* (22 November 1913), p. 5.
14  *Sydney Morning Herald* (20 September 1917), p. 6; *Evening News* (19 September 1917), p. 5.
15  The 'tram loop' at the north-eastern corner of the park was managed by the Railway Commissioners not the Council, meaning carnivals continued in that corner of Bondi: Patricia Quinn-Boas, 'Bondi 1920–1940: The Development of an Urban Recreation Area by Waverley Municipal Council' (unpublished Master's thesis, University of Sydney, 1967), p. 16.
16  Curby, *Seven Miles From Sydney*.
17  SRNSW: NRS8258 [33/7088].
18  Ibid.
19  *Sydney Morning Herald* (18 December 1934), p. 11; (26 December 1964), p. 14; (19 February 1935), p. 9.
20  The Council had been quick to deny the application, most likely due to the heavy investment they were making into the Coogee Pier, itself designed to attract those seeking amusements on the foreshore: *The Sun* (9 July 1925).
21  The ex-Mayor denied this: SRNSW NRS8258: 33 7088; *Sydney Morning Herald* (8 January 1932), p. 10.
22  Sue Rosen, *Historical Outline: Luna Park/Lavender Bay Heritage Study* (Sydney: Godden MacKay for NSW Department of Planning, 1991), pp. 37–41.
23  *Sydney Morning Herald* (5 October 1935), p. 18.
24  *Eastern Free Press* (10 April 1930), p. 1.
25  SRNSW: NRS8258 [33/7088]; *Sydney Morning Herald* (19 December 1934), p. 14.
26  Peggy James, *Cosmopolitan Conservationists: Greening Modern Sydney* (Melbourne: Australian Scholarly Publishing, 2013), pp. 214–15.
27  *Sydney Morning Herald* (26 December 1934), p. 5; *Truth* (11 November 1929), p. 1.
28  SRNSW: NRS8258 [33/7088].
29  Justin de Gouw, 'Luna Park: Just for Profit!' (unpublished B. Arch. dissertation, University of New South Wales, 1991), p. 2.
30  de Gouw, 'Luna Park', p. 15; SRNSW: NRS8258 [33/7088].
31  Leroy Ashby, *With Amusement For All: A History of American Popular Culture Since 1830* (Lexington: University Press of Kentucky, 2006), p. 107; Cross and Walton, *The Playful Crowd*, pp. 97–8.
32  *Sydney Morning Herald* (19 December 1934), p. 14.
33  SRNSW: NRS8258 [33/7088].
34  Evidence to Metropolitan Land Board Inquiry 1932–3, SRNSW: NRS8258 [33/7088].
35  Cultural attachment to trees is not unique to Bondi. Anthropologist Setha Low described a number of 'occupants' of Costa Rica's *parque central* who wept at the loss of two giant palm trees from the plaza. An old man described them as 'like friends (that) made his bench a special place': Setha Low, 'Symbolic Ties That Bind', in Setha Low and Irwin Altman (eds), *Place Attachment* (New York: Plenum Press, 1992), p. 179.
36  James, *Cosmopolitan Conservationists*, pp. 112, 215; SRNSW: NRS8258 [33/7088].
37  Caroline Ford, 'The First Wave: The Making of a Beach Culture in Sydney, 1810–1920' (unpublished Doctoral thesis, University of Sydney, 2007).
38  SRNSW: NRS8258 [33/7088].
39  Pauline Curby, *Randwick* (Sydney: Randwick City Council, 2009), p. 261.
40  Brawley, *The Bondi Lifesaver*, p. 99.

41  Ford, *Sydney Beaches*.

42  SRNSW: NRS8258 [33/7088].

43  de Gouw, 'Luna Park', p. 14; *Sydney Morning Herald* (8 April 1935), p. 8.

44  Ian Hoskins, *Sydney Harbour: A History* (Sydney: UNSW Press, 2009), p. 224.

45  de Gouw, 'Luna Park', pp. 5, 15.

46  North Sydney Council Minutes (27 July 1937).

47  Rosen, *Historical Outline*, pp. 47, 50.

48  Rosen, *Historical Outline*, pp. 47, 69.

49  Friends of Luna Park, *Report on Luna Park, Sydney, Australia* (1980).

50  Yi-Fu Tuan, *Topophilia: A Study of Environmental Perception, Attitudes and Values* (Englewood Cliffs: Prentice-Hall, 1974), pp. 99, 197.

51  Godden MacKay Pty Ltd, *Luna Park/Lavender Bay Heritage Study: Volume One* (Prepared for the NSW Department of Planning, 1991), p. 11.

52  http://www.environment.nsw.gov.au/heritageapp/ViewHeritageItemDetails.aspx?ID= 5055526 accessed 20 June 2014.

53  See, for example, Jan Roberts (ed.), *Remembering Avalon: Growing up in the 1940s and '50s* (Avalon Beach: Ruskin Rowe Press, 2011), pp. 113–14.

54  Douglas Booth, 'Bondi Park: Making, Practicing and Performing a Museum', in Murray G. Phillips (ed.), *Representing the Sporting Past in Museums and Halls of Fame* (London: Routledge, 2013), pp. 204–30.

55  Anthropologist Setha Low describes cultural places attachment at a complex web of relationships that people create with places through multiple linkages that include emotions, practices and cultural beliefs: Low, 'Symbolic Ties That Bind', p. 165.

# 5 The Turkish amusement park

## Modernity, identity and cultural change in the early Republic

*Jason Wood and B. Nilgün Öz*

The conception of the Turkish Republic in 1923 as a secular, modern nation-state, distant from the preceding Ottoman Empire, heralded a seminal period of radical reform that saw the adoption of western attitudes, styles and institutions most conspicuously in the spheres of politics, education and law and, importantly in this context, social and cultural life. As a result, this modernist project of social engineering and cultural change laid the foundations for a new Turkish national identity.[1]

In urban contexts, the creation and role of parks, sports facilities and other places of public recreation and entertainment were significant conduits of socio-cultural transformation and important symbols of early Republican modernity and inclusivity.[2] Bringing men and women into close proximity in new public spaces worked to destabilise the Islamic tradition of gender segregation, while the places themselves allowed people to practise their new Turkish identities and imbue them with nationalism.[3]

Whilst acknowledging the specific impact the urban park movement had on the proliferation and permanency of leisure facilities and, as will be shown, on amusement parks in particular, it would be amiss to suppose that mixed-gender entertainment was a new social experience at this time. The transition from the late Ottoman to early Republican periods saw a gradual rise in the use of dedicated, although temporary, places for entertainment in numerous towns and city districts across Turkey. Predominantly restricted to holy days – known as *bayrams* – and other official holidays, this entertainment took the form of travelling fairs and circus acts, including a variety of acrobats, magicians, puppeteers, sword-swallowers and fire-eaters; donkey rides and wild animal shows; and early amusement rides such as wooden swinging boats, revolving chairs (with separate ones for children and adults), roundabouts, carousels, and even rudimentary dodgem cars and 'fear tunnels' (a reference to ghost trains) (Figure 5.1).[4] It was in these temporary home-grown installations, and in later foreign travelling circuses and funfairs, that the Turkish amusement park had its origin.

This short chapter will explore the creation and development of amusement parks in Izmir and Ankara, in the respective contexts of Izmir's Culture Park and Ankara's Youth Park, and within the larger framework of modernity, identity and cultural change in Turkey during the first decades of the nation-building process.

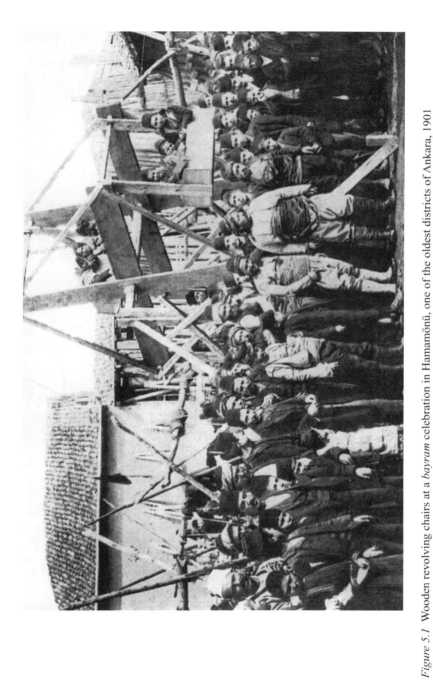

*Figure 5.1* Wooden revolving chairs at a *bayram* celebration in Hamamönü, one of the oldest districts of Ankara, 1901

Source: Moughamian Freres postcard © Koç University VEKAM Archive (0989)

In different ways, Izmir and Ankara were pioneering cities whose urban parks served as instruments of the new Republican regime and its ideology and as significant agents of transformation in the public realm and social lifestyles. Their eventual role, however, in propagating new entertainment cultures, routines and practices, by providing enduring homes for what became two of Turkey's earliest Luna Parks, is a neglected subject that has given rise to some misconstrued thinking.

## Izmir

As a port city and tourist destination, Smyrna – modern-day Izmir – developed an increasingly cosmopolitan character and westernised lifestyle throughout the late Ottoman period. It featured in Baedeker guidebooks and Murray's handbooks and welcomed foreign visitors arriving daily on passenger liners. The Grand Hotel Kraemer Palace hosted Thomas Cook tours, while cinema was embraced as early as 1908. Tragically, however, by the time the Republic was proclaimed, Izmir was a city in ruins following a devastating fire that started on 13 September 1922, destroying over 300 hectares including the central business district and residential areas.[5]

The years between 1923 and 1950 saw the city reconstructed, the economy revived and social life reorganised.[6] The most striking consequence of this development was the creation, in the fire zone, of a Culture Park (*Kültürpark*) completed in 1936. The new Turkish state saw Culture Parks as supporting cultural enlightenment for its modern citizens. Earlier, in an attempt to encourage growth and prosperity, Mustafa Kemal (Atatürk) had suggested that Izmir become a 'city of trade fairs'. The Culture Park quickly became the location for Izmir's world-renowned International Fair as well as other economic and recreational purposes throughout the year, including a small funfair that grew to become the city's Luna Park.[7]

The idea to create a Culture Park derived from the strong will and single-minded determination of Dr Behçet Uz, mayor of Izmir (1931–41) while the initial proposals were inspired by a visit in 1933 to Gorky Park in Moscow.[8] Following the visit, and at the instigation of the Izmir Municipality, a report was prepared in 1934 by Suat Yurdkoru, head of Izmir Football Contingency, recommending the various types of facilities that the Culture Park might contain.[9] In Yurdkoru's report it was claimed that 'the Culture Park [had] been designed to fulfil the need of outdoor pursuits for Izmir dwellers and to appeal to the aesthetic taste'. The report went on to say that 'This facility should be used by the new generations to achieve the cultural tasks of the revolution'.[10] In presenting the case to the Municipality, before seeking final approval from the Turkish General Assembly, mayor Uz envisioned the Culture Park as 'a people's university . . . a green zone in which the public's cultural needs will be met'.[11] 'The public', he said, 'should be able to breathe, get to know certain plants, and gather information on how to raise children, how a healthy human being develops, and has fun'.[12]

The central location chosen for the development of the Culture Park covered an important area of the fire-damaged city, and one of the foci of the masterplan prepared by French urban planners René and Raymond Danger, in collaboration with Henri Prost, for the city's reconstruction. Although completed in 1924–5, the masterplan remained in effect throughout 1930s, the layout of some of the roads in particular shaping the morphology of the Culture Park.[13] From 1934, the fire debris was cleared, and over the next two years Izmir's Culture Park emerged amidst the ruins, like the oasis imagined by mayor Uz. On 1 September 1936, it was officially opened by the Prime Minister (and later second President of the Republic), Ismet Inönü.

Planned in the shape of an irregular hexagon, dissected by wide, tree-lined boulevards and entered through five monumental gates, the Culture Park covered an immense area of 360,000 square metres. This was subsequently expanded in 1938–9 to 420,000 to accommodate the growing needs of the International Fair. As well as the main exhibition palace and the exhibition pavilions of numerous countries, Turkish provinces and organisations, the Culture Park comprised ornamental lakes adorned with statues and artworks; canoeing, boating and swimming pools; various museums (including a Museum of Health, later converted to an Archaeology Museum); an open-air theatre (featuring everything from dramas to wrestling); a Parachute Tower (the tallest structure in Izmir at that time and intended to introduce youngsters to aviation through parachute jumping); a zoo; children's playground; and an assortment of clubs, restaurants, outdoor cafes and tea gardens (Figure 5.2). The architecture of all the Culture Park buildings was in the modernist style.[14]

The history and development of the Luna Park are sketchy.[15] Its origins would appear to lie in a foreign travelling circus. In Yurdkoru's 1934 report, prepared after the visit to Gorky Park, provision was made for the Culture Park to incorporate open-air recreational facilities, including a *sirk* or 'circus place'. In September 1936, after the Culture Park opened, the Kluski Circus (*Kluski Sirki*) came to the International Fair. Arriving in forty large wagons, it was the first time this circus had travelled to Turkey. The show, which ran twice daily throughout the duration of the Fair, featured acrobats and one hundred wild animals.[16]

The Luna Park was probably established soon after this, initially as a small funfair. It certainly existed by 1938, attested in various photographs and newspaper accounts. Photographs taken from the top of the Parachute Tower show its location at this date on the southern edge of the Culture Park, close to the site of the large, new Soviet Pavilion which was built when the Culture Park was expanded by 60,000 square metres in 1938–9 (Figures 5.3 and 5.4).[17] As well as a circus 'big top', amusements included something called a 'Hilarity Shack' (probably a fun house), 'Satan's Box', a Hall of Mirrors, at least two 'Wall of Death' motorcycle rides, various swings, Flying Chairs, and a Big Wheel (Figure 5.5).[18]

The Big Wheel was the star attraction: 'The fun that old, mature men and women are having by sitting in the big wheel [written as 'the big cupboard'] is one of the must-sees of the Fair'.[19] The ride was not to everyone's liking, however, as the following episode related by the well-known journalist and cartoonist Orhan Ramiz Gökçe shows: 'Amazing laughter comes from further on. I move towards the middle. A wheel, a giant wheel! . . . And an old woman in one section of the wheel. She is screaming her lungs out: Something's come over me, please let me

*Figure 5.2* An aerial view of the Culture Park, 1958

Source: Turizm Fotoğraf Ajansı © The authors' collection

*Figure 5.3* The small funfair on the southern edge of the Culture Park, before the Soviet Pavilion was built, viewed from the top of the Parachute Tower, about 1938

Source: Foto Cemal © Izmir Chamber of Commerce Glass Plate Negative Collection

*Figure 5.4* The expanded Luna Park, after completion of the Soviet Pavilion, viewed from the same vantage point, about 1940

Source: Foto Cemal © Izmir Chamber of Commerce Glass Plate Negative Collection

*Figure 5.5* The Flying Chairs and Big Wheel

Source: Foto Cemal © Izmir Chamber of Commerce Glass Plate Negative Collection

get off . . .'. Gökçe's visit was in August 1938 and, other than the old woman's misadventure, he paints a picture of a lively, happy and contemporary place where men and women mingle together and enjoy life.[20] A new ride created in 1939 also won favour. Called the 'Live Duck', but actually consisting artificial swans, they 'take their customers on board and then take off using electricity and amuse the visitors to the Fair' (Figures 5.6 and 5.7).[21]

The Luna Park was forced to relocate to its current site in the south-east corner of the Culture Park when a new set of exhibition halls were erected to accommodate the growing demands of the International Fair. None of the historic rides were retained.[22]

## Ankara

Despite being made capital of the new Turkish state in 1923, Ankara offered its citizens limited opportunities by way of leisure and amusement in the early years of the Republic. Social life was restricted to a few district parks and restaurants, and only two cinemas. In the late 1920s and early 1930s, people had to travel some distance from the city centre to satisfy their recreational needs. The principal destinations at this time were the Atatürk Orman Çiftliği, or Model Farm and Forest, where boating and swimming pools, and a zoo, were available from 1925; and the Çubuk Dam Lake where boating could be enjoyed from 1929.[23]

It would be several more years before the opening of the Youth Park (*Gençlik Parkı*), the first and largest urban park in the heart of Ankara and indeed one of the most important open public spaces created by the Republican regime. Youth Parks were intended to satisfy popular demand for leisure and recreational activities and to foster new generations of 'young' and 'healthy' citizens. Known as the 'Heaven of the Steppes', Ankara's Youth Park quickly became a symbol of the Republic itself, its identity and the new capital city.[24]

An international competition to develop a masterplan for Ankara was won by the German city planner Hermann Jansen in 1927. Jansen's first design proposals for the Youth Park were conceived in 1928, and he continued to work on them until his contract was terminated in 1935. Thereafter, plans for the Youth Park were modified by the French landscape architect and planner Theo Leveau, and these changes were approved in 1936. Construction finally began in 1938 but it would be five years before the Youth Park was entirely completed. It was officially opened, in the presence of the President of the Republic Ismet İnönü, on 19 May 1943, as part of celebrations to mark national Youth and Sports Day. During the 1950s, the Youth Park acquired a small funfair and miniature railway. The popularity of these additional attractions grew, boosted by visitors to the Ankara Exhibition (*Bugünkü Ankara*), and by the end of the decade the Youth Park became the permanent home for the city's Luna Park.[25]

The site chosen for the Youth Park lay immediately adjacent to Ankara's sports stadium and racecourse, both designed by the architect Paolo Vietti-Violi and opened in 1936, and a Parachute Tower completed in 1937, similar to the one in Izmir. Laid out on reclaimed marshland, between the city's old district

*Figure 5.6* The 'Live Duck' ride and Big Wheel

Source: Foto Cemal © The authors' collection

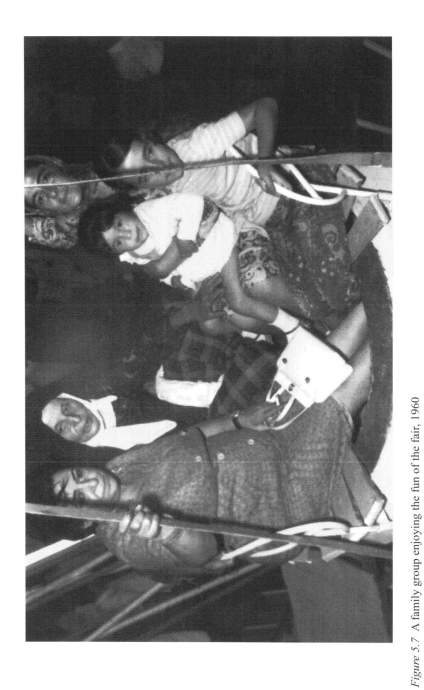

*Figure 5.7* A family group enjoying the fun of the fair, 1960

Source: © The authors' collection

and the railway station, the Youth Park covered an area of 280,000 square metres and was roughly triangular in shape, with originally three gates. The influence of Leveau is apparent in that its geometry and symmetry resembled a French formal garden. The main feature was a large artificial lake (42,000 square metres) with an ornamental fountain (Figure 5.8). The lake became a favourite place for water sports (swimming, rowing, boating and sailing) and also ice-skating in winter. Around this were various sports and games areas; a running track; a children's playground; an open-air theatre (for musical and theatrical performances); and various restaurants, coffee houses and tea gardens. The modernist style is reflected in the architecture of all the Youth Park buildings.[26]

As for the origins and development of the Luna Park, it is known that Jansen had planned an 'entertainment zone' in the south-west area of the Youth Park and Leveau's plan appeared to follow suit.[27] When the Youth Park opened, however, 'completion of [the] children's playground, restaurant, Lunapark, and the

*Figure 5.8* An aerial view of the Youth Park and Ankara's sports stadium and racecourse beyond, 1953

Source: © Koç University VEKAM Archive (0523)

open-air theatre were left to the future because of the restrictions in construction materials'.[28] Several years later, in 1951, a travelling funfair, promoted by an Italian company that had previously visited Istanbul, was given space in the Youth Park (Figure 5.9). Its success and acclaim seem to have been the final stimulus for the Youth Park authorities to create a funfair of their own. This opened in 1952, in the south-east area of the Youth Park, on land apparently set aside for the children's playground, and became a regular, although at this time still only temporary, attraction.[29]

A postcard of 1955, taken from across the lake, shows what by this date was known as the Luna Park. Amusements apparently included a Big Wheel, Flying Chairs and 'Wall of Death' motorcycle rides (Figure 5.10).[30] In another undated photograph a racing car ride is shown. The well-known author Buket Uzuner, writing about her childhood in 1955, fondly remembers the Big Wheel and also a carousel:

> When it got dark, the sparklingly bright Luna Park would turn into a carnival for the children who would fling themselves to its arms with irresistible love, and the evening would end with the carousel, ridden with joyful cries, and the pleasure of the wheel.[31]

By 1956 the Luna Park had acquired new rides. A mini-golf course was also created close by. These changes coincided with the opening of the Ankara Exhibition, which ran from 1956 to 1958 and was housed in a building just outside the southeast area of the Youth Park (now the Ankara Opera House). The Luna Park was clearly sufficiently well known during this period as a disagreement between the Luna Park manager and the exhibition organisers made headline news.[32] In 1957, a miniature railway was installed to carry sightseers and the less able on journeys around the perimeter of the Youth Park. Two trains, 'Mehmetçik' and 'Efe', provided by Turkish State Railways, operated along a 1750 metre long track, stopping at four stations, including one next to the Luna Park.[33]

After the closure of the Ankara Exhibition at the end of 1958, a new plan (dated 1959) was approved for the reorganisation of the Youth Park. This led to a decision to move the Luna Park to the much larger 'entertainment zone' in the south-west area of the Youth Park, as Jansen/Leveau had originally designated it. Later, a wall was built to separate the Luna Park from the Youth Park, since when the area has been the Luna Park's permanent home.[34]

## Discussion

Izmir's Culture Park and Ankara's Youth Park were both large areas in central locations laid out as part of city-wide reconstruction projects. Both were characterised by geometric shapes, regularised landscape design and modernist architecture. They also changed the scenery by introducing trees, gardens and rare bodies of water into an urban setting. What makes the Culture Park and Youth Park special among other public parks or district parks in Izmir and Ankara, or

*Figure 5.9* Riding a dodgem car at the Italian funfair in the Youth Park. The photograph was sent by Mahmut Z. Zelcan to his aunt and dated 27 September 1951

Source: © The authors' collection

*Figure 5.10* A view from across the lake showing the temporary location of the Luna Park including a Big Wheel, Flying Chairs and 'Wall of Death' motorcycle rides, about 1955

Source: © Koç University VEKAM Archive (0557)

indeed elsewhere in Turkey, are the ideological and symbolic meanings attributed to them. The reclaiming of the fire zone in Izmir and the marshland in Ankara can be read as powerful metaphors and emblematic acts of socio-cultural change – from fire-blackened ruins to green, cultural oasis; from mosquito-infested swamp to healthy recreational lake; from Ottoman Empire to Turkish Republic. Offering many year-round cultural, sporting and leisure opportunities, the Culture Park and Youth Park operated as media through which contemporary Republican citizens, from different social classes, and both men and women, could live out their transforming social and gendered identities and liberate themselves from the practices of the traditional Ottoman society they were replacing.[35] Additionally, the Culture Park from the outset in 1936, and the Youth Park some twenty years later, became places used for economic as well as socio-cultural purposes. As the location of the Izmir International Fair, the Culture Park was host to an important commercial event not just for the city but for the country. Similarly, in the Youth Park, the Ankara Exhibition, although relatively short-lived, was also a significant economic success and a destination for visitors from further afield.

The Izmir and Ankara Luna Parks owe their respective origins to a foreign travelling circus and a foreign travelling funfair. It seems clear that Izmir's Luna Park secured a permanent location in the Culture Park, at some date between 1936 and 1938, due to its direct association with the international trade fairs from which it gained the necessary impetus and clientele. The situation in Ankara is more puzzling. Here the Luna Park, founded in 1952, existed, albeit temporarily, for four years before it benefitted from the capital city's major exhibition in 1956–8. From 1952 to when the Luna Park achieved permanency in 1959, the sources are relatively silent. The Luna Park's historical and continuing presence in the Youth Park has, however, become something of a one-sided argument, which is long overdue for correction.

The debate starts from a false premise that the Luna Park was equated with what is termed a 'radical change' in the history of the Youth Park brought about by the establishment of the Ankara Exhibition. In the key references, words and phrases like 'inserted' or 'invaded by' are used to describe the Luna Park's appearance in the Youth Park.[36] The Ankara Exhibition, Luna Park and miniature railway are seen as arbitrary 'interventions' marking the beginning of a 'disintegration' or 'fragmentation process', by manifestly and decisively deviating from, and undermining, the Youth Park's original Republican values and identity.[37] The new emphasis on commercialism and entertainment, at the expense of the 'alluring image' of Republican citizens in calm repose in the greenery of the Youth Park, is portrayed as disregarding, over-dominant 'Disneyfication'.[38] This theme is repeated in what amounts to Ankara's official history which sees the 'degeneration' of the Youth Park starting even earlier, in 1951, with the arrival of the Italian funfair. The authors describe a carnival atmosphere, more like a 'continuous festival . . . with tea gardens turned into alcoholic gazinos meaning that local Ankara families ceased going to the Youth Park, leaving the area to more-moneyed but lower social classes'.[39]

Yet, according to the first of the key references for the Youth Park, the designation of an 'entertainment zone' and the creation of a Luna Park were in the original

Jansen/Leveau proposals, and only delayed due to lack of building materials.[40] So, why is the provision of an 'entertainment zone', and the Luna Park in particular, rendered so negatively when they were both planned from the start? And why deploy the emotive degenerative comparison to refer to the emergence of the new consumer society from the 1950s, when the Youth Park had always been popular among different social groups and, in fact, inclusively designed to be so?

To explain such contradictions and stereotyping it is necessary to look further than the inconsequential argument that the 'insertion' of the Luna Park on land earmarked for the children's playground disenfranchised a number of children and their parents.[41] Further, too, than the obvious lack of literature on the development of the Turkish amusement park, although comparative studies do exist for other forms of leisure and entertainment in the early Republic.[42] The answer lies, it is suspected, in an understanding and appreciation of the social, political and academic contexts in which the authors were writing.

First, the social context. From the 1970s, the Youth Park was in decline and its infrastructure deteriorating. The lake was no longer used for water sports and in general the Youth Park was beginning to lose its attractiveness, popularity and meaning among the citizens of Ankara. This was primarily the result of neglect by the relevant authorities, which led to an escalation in use by migrants and low-income groups like the homeless and unemployed. How the Luna Park was affected by the decline of the Youth Park is unknown, although it had by now acquired the wall separating it from the Youth Park and was beginning to operate semi-independently.[43] Could it be that the authors or their sources have transposed these societal changes from the 1970s to their interpretation of events in the 1950s, finding a convenient root cause in the shape of the Luna Park?

Second, the political context. Two instances, relating to the Youth Park and the fate of its Luna Park, will suffice to help explicate the problem. In 1981 it was declared that a new Atatürk Cultural Centre would be built in Ankara, to commemorate the centenary of Atatürk's birth. As part of this development plan, a new educational science park was to be located in the Youth Park, involving the demolition of the Luna Park 'along with other structures that did not merit conservation'. Historiographically, this plan is presented as a 'turning point' and an attempt to recapture the original social meaning and cultural identity of the Youth Park.[44] In the end, the proposals, as far as they affected the Youth Park, never materialised.[45]

In much the same vein, a second 'turning point' was reached in 2005. By this date the Youth Park had become very run down indeed.[46] Plans were therefore submitted for its regeneration as part of a re-evaluation of the Atatürk Cultural Centre. In a report by the Chamber of Architects Ankara Office for the Greater Ankara Municipality (the Youth Park's owners), the removal of the Luna Park was again proposed as it, and its immediate setting, were seen to have 'damaged the main aim of the Youth Park, following a number of wrongheaded planning applications'.[47] Unsurprisingly, the owners of the Luna Park objected and the regeneration project stalled for legal reasons. Eventually, it was agreed that the Luna Park could stay with only minor modifications. Paradoxically, in an intervention by the Mayor of Greater Ankara Municipality, nostalgia for the miniature train, 'swept away' in the 1990s, was invoked to support the idea (which came to nothing) of installing a monorail in the Youth Park.[48] The regenerated Youth Park reopened in 2009. According to the architect, Öner Tokcan, the main aim was to give the Youth Park a 'modern' facelift without damaging its historic texture. In reality, this was a *de novo* proposal that 'destroyed its aesthetics and functionality'.[49] New buildings and structures erected between 2009 and 2012 introduced further contradictions and political tensions concerning the provocative choice of Seljuk and Ottoman architectural styles. Improvements to the Luna Park, including the addition of new rides to appeal to a wider audience, were also made at this time.[50]

Could these political interventions, seeking but ultimately failing to remove the Luna Park, and giving rise to a protracted period of uncertainty from 1981 to 2005, have played on the minds of the authors and fixed them with the notion that differentiated the Youth Park as good, worthy of investment and rebirth; from the Luna Park as bad, unmerited and damaging?

Third, and finally, the academic context. Ankara is an especial case in this regard, with two distinct historiographies being prevalent from the late 1990s:

> studies aiming to uncover the relations between the social dynamics of the early Republican period and urban space; and . . . a growing literature that reproduced a nostalgic image of this period, especially of Ankara as the symbolic locus of [Turkish modernisation].[51]

The key references for the Youth Park fall within this period (1998–2012) so both of these historiographies are at play here. Prevailing also, particularly in the theses, is a habitual and deferential use of traditional sources relating to the history of Turkish nationhood and modernity.[52] But it seems there was no attempt, like some scholars and intellectuals, to look at modernisation from alternative, pluralist and 'unofficial' vantage points, to 'favor and emphasize cultural identity, differences, and diversity over homogenization and unity'.[53] This gets to the nub. The authors have perhaps been reluctant to engage with public history, popular culture and the unconventional, but as architects and city planners as most of them are, is it that they have been conditioned only to crave design, order and regulation? The Luna Park, of course, was anything but.

## Conclusion

The nascent Turkish Republic gave rise to a modern society out of a traditional one. The construction of a new national identity was underpinned by a total socio-cultural transformation that empowered notions of secularity, class and gender equality, and embraced western styles of living. The creation of public spaces in the urban environment, like Izmir's Culture Park and Ankara's Youth Park, brought people together to express their modernity and identity, and to signify their aspiration to belong to a wider world. Enjoying casual and pleasurable activities, such as going to an open-air theatre, taking in a museum, enjoying water sports, or even jumping off a Parachute Tower, became a significant part of this newly adopted lifestyle.

If urban parks were landmarks of modernism, and schools for socialising people, so, too, were the amusement parks that grew and prospered within their economically fertile confines. Rightly repositioned in the case of Ankara's Luna Park, these should now be seen as contributing to, rather than detracting from, the so-called 'golden age' of early Republican modernity, identity and cultural change.

## Notes

1  For this theme see Hale Yılmaz, *Becoming Turkish: Nationalist Reforms and Cultural Negotiations in Early Republican Turkey (1923–1945)* (Syracuse University Press, 2013). Yılmaz offers a rare perspective that stresses social and cultural dimensions and everyday negotiations of the reforms. See also Ayhan Akman, 'Ambiguities of Modernist Nationalism: Architectural Culture and Nation-building in Early Republican Turkey', *Turkish Studies* 5, 3 (2004), pp. 103–11 (esp. p. 110).

2  Sibel Bozdoğan, *Modernism and Nation Building: Turkish Architectural Culture in the Early Republic* (Seattle: University of Washington Press, 2001), pp. 75–9.

3  Ayşe Burçin Erarslan, 'Simultaneous Projects of Nation-building and City-building: Three National Spatialities in Izmir, Ankara, Istanbul'. *Paper presented at the American Sociological Association Annual Meeting, Hilton Atlanta and Atlanta Marriott Marquis, Atlanta, GA* (14 August 2010), pp. 7–9: http://citation.allacademic.com/meta/p_mla_apa_research_citation/4/0/8/8/1/pages408815/p408815-1.php accessed 13 September 2014.

4  R. Sertaç Kayserilioğlu, 'Tarih olan bayram yerleri', *#tarih* 3 (August 2014), pp. 55–62. Kayserilioğlu is a researcher, writer and broadcaster with a substantial collection of postcards and photographs: http://www.sertackayserilioglu.com/ accessed 13 September 2014. The article is illustrated using images from his collection, and the archive of Cengiz Kahraman, and includes photographs from Trabzon (as early as 1924), Ankara (1940), Istanbul, Manisa and Edirne. See also Ozan Sağdıç, *Bir Zamanlar Ankara* (Ankara Büyükşehir Belediyesi, 1993), pp. 100–101.

5  The best account of popular entertainment in Izmir in the mid to late nineteenth century is Rauf Beyru, *19. Yüzyılda Izmir Kenti* (Istanbul: Literatür, 2011), pp. 225–41. See also Giles Milton, *Paradise Lost. Smyrna 1922: The Destruction of Islam's City of Tolerance* (London: Hodder & Stoughton, 2008) for a compelling account of Smyrna in the first two decades of the twentieth century before the fire.

6  For a pictorial history of this period see Erkan Serçe, Fikret Yılmaz and Sabri Yetkin, *Küllerinden Doğan Şehir/The City which rose from the Ashes* (Izelman Ltd, 2003) and Fikret Yılmaz (ed.), *History Written on Glass* (Izmir Chamber of Commerce Glass Plate Negative Collection, 2007). Both are copiously illustrated with images from 1922 onwards, mostly by Cemal Yalkış (1903–2001), known as 'Foto Cemal'.

7  For general background on the Culture Park and International Fair see Anon, *70 Yıllık Sevda: Izmir Fuarı* (Izmir Büyükşehir Belediyesi, 2001) and Elvan Feyzioğlu, *Büyük Bir Halk Okulu Izmir Fuarı* (IZFAŞ, 2006). The International Fair, the oldest trade fair in Turkey, is known as the 9 Eylül Panayırı (9 September Fair) after the date of the Izmir's liberation in 1922 at the end of the War of Independence. It was inaugurated in 1927 and held annually until 1933, before moving to its purpose-built venue in the Culture Park in 1936. The Fair continues to be a permanent fixture in the park despite the opening of a new 'Fuar Izmir' complex close to the airport.

8  Didier Laroche, 'Izmir et le Débat Architectural en Turquie dans les Années 30 et 40, in *Le Corbusier en Turquie: Le Plan Directeur d'Izmir (1939–1949)* (Strasbourg: École Nationale Supérieure d'Architecture de Strasbourg, 2009), pp. 24–39 (p. 31).

9  For the development and various components of the Cuture Park see Serçe et al., *Küllerinden Doğan Şehir*, ch. 6; Yılmaz, *History Written on Glass*, ch. 7; and Anon, *Üç Izmir* (Yapı Kredi Yayınları, 1992), pp. 314–20.

10 Serçe et al., *Küllerinden Doğan Şehir*, pp. 158–9. See also the accounts of Yurdkoru in Feyzioğlu, *Bûyûk Bir Halk Okulu Izmir Fuarı*, pp. 29–31.

11 Serçe et al, *Küllerinden Doğan Şehir*, pp. 158–60.

12 Anon, *Üç Izmir*, p. 315.

13 K. Cana Bilsel, 'Ideology and Urbanism during the Early Republican Period: Two Master Plans for Izmir and Scenarios of Modernization', *METU JFA* 16, 1–2 (1996), pp. 13–30 (p. 14): http://jfa.arch.metu.edu.tr/archive/0258-5316/1996/cilt16/sayi_1_2/13-30.pdf accessed 15 September 2014. In 1938, the Municipality approached Le Corbusier to prepare a new masterplan for Izmir but this was eventually judged unfeasible: see *Le Corbusier en Turquie* (see n. 8).

14 Bozdoğan, *Modernism and Nation Building*, pp. 130, 141–7. For a plan of the Culture Park in 1936 see Serçe et al, *Küllerinden Doğan Şehir*, p. 174. For an illustrated architectural guide to what still survives from the late 1930s see Anon, *Izmir Mimarlık Rehberi 2005* (Mimarlar Odası Izmir Şubesi, 2005), pp. 128–33, which features the Museum of Health (1937), Parachute Tower (1937), Pakistan Pavilion (1938) and Cultural Pavilion (1939). See also Meltem Ö. Gürel, 'Architectural Mimicry, Spaces of Modernity: The Island Casino, Izmir, Turkey', *The Journal of Architecture* 16, 2 (2011), pp. 165–90.

15 With the notable exception of Feyzioğlu, *Bûyûk Bir Halk Okulu Izmir Fuarı*, there is virtually no mention of the Luna Park in any of the literature relating to the Culture Park or International Fair.

16 Anon, *Üç Izmir*, p. 315; Feyzioğlu, *Büyük Bir Halk Okulu Izmir Fuarı, p. 43.

17 Serçe et al., *Küllerinden Doğan Şehir*, p. 190; Yılmaz, *History Written on Glass*, pp. 222–3.

18 Serçe et al., *Küllerinden Doğan Şehir*, p. 198; Yılmaz, *History Written on Glass*, pp. 256–7; Anon, *Kalbim Ege'de Kaldı: Cemal Yalkış'ın gözüyle bir zamanlar Izmir* (Istanbul: E.G.S. Bank, 1998), pp. 104, 109, 111.

19 Newspaper article quoted in Feyzioğlu, *Bûyûk Bir Halk Okulu Izmir Fuarı*, p. 54.

20 Orhan Ramiz Gökçe, 'Izmir Fuarında Bir Gece', *Cumhuriyet* (31 August 1938), quoted in Feyzioğlu, *Büyük Bir Halk Okulu Izmir Fuarı*, pp. 52–3.

21 Newspaper article quoted in Feyzioğlu, *Büyük Bir Halk Okulu Izmir Fuarı*, p. 61; Yılmaz, *History Written on Glass*, p. 256.

22 For contemporary images of the park today see https://www.flickr.com/photos/cerased/4914177862/in/photostream/ accessed 15 September 2014.

23 Inci Aslanoğlu, '1930–1950 Yılları Ankara'sının Eğlence Yaşamı İçinde Gazino Binaları', in Yıldırım Yavuz (ed.), *Tarih İçinde Ankara II* (Ankara: ODTÜ Mimarlık Fakültesi, 2001) pp. 327–9.

24 Bozdoğan, *Modernism and Nation Building*, pp. 70, 75. For this theme see Bülent Batuman, '"Early Republican Ankara": Struggle over Historical Representation and the Politics of Urban Historiography', *Journal of Urban History* 37, 5 (2011), pp. 661–7: http://yoksis.bilkent.edu.tr/pdf/files/10.1177-0096144211407738.pdf accessed 27 September 2014.

25 The key references for the development of the Youth Park and its changing meanings and values are three unpublished theses (one PhD and two Masters) from the Middle East Technical University, Ankara: Zeynep Sökmen Uludağ, 'The Social Construction of Meaning in Landscape Architecture: A Case Study of Gençlik Parkı in Ankara' (Doctor of Philosophy in Architecture, METU, 1998); Can Akansel, 'Revealing the Values of a Republican Park: Gençlik Parkı Deciphered in Memory and as Monument' (Master of Architecture in Architecture, METU, 2009); and Nihan Oya Memlük, 'Inclusivity of Public Space: Changing Inclusivity of an Urban Park, Gençlik Parkı, Ankara' (Master of Science in City and Regional Planning, METU, 2012). See also Erol Demir, 'Toplumsal Değişme Süreci İçinde Gençlik Parkı: Sosyolojik Bir Değerlendirme', *PLANLAMA* 4 (2006), pp. 69–77, who is particularly good on the evolving physical structure of the park and social composition of its users.

26  Bozdoğan, *Modernism and Nation Building*, pp. 75–7; Akansel, 'Revealing the Values of a Republican Park', pp. 71, 99. For images of the Youth Park see Atila Cangır (ed.), *Cumhuriyetin Başkenti*, 3 vols (Ankara Üniversitesi, 2007), pp. 793–4 (under construction in 1940), p. 1047 (in 1943), p. 1050 (in 1945), pp. 809, 819 (in 1950), pp. 816, 821, 918–9, 962, 1047–8 (in 1955), pp. 1030, 1049 (in 1960), p. 1029 (in 1965). See also Anon, *Fotoğraflarla Dünden Bugüne Ankara. Past and Present by Ankara in Photographs* (Ankara: T.C. Millî Eğitim Bakanlığı Eğitim Teknolojileri Genel Müdürlüğü, 2006), p. 52 (in 1949), pp. 55–7 (in 1953), pp. 30–31 (1960s and 70s).

27  Akansel, 'Revealing the Values of a Republican Park', pp. 63, 65, 70.

28  Uludağ, 'The Social Construction of Meaning in Landscape Architecture', p. 182.

29  Ibid., p. 204; Akansel, 'Revealing the Values of a Republican Park', pp. 105–6; Memlük, 'Inclusivity of Public Space', pp. 92–3.

30  Cangır, *Cumhuriyetin Başkenti*, p. 821. The postcard was originally published by Editörü ve Fotoğrafçısı Bilinmeyen Kartlar (VEKAM Ankara Collection 1339). Another photograph of a similar view is published in Anon, *Fotoğraflarla Dünden Bugüne Ankara*, p. 30 but is wrongly dated to 1960 (VEKAM Ankara Collection 0557 – again wrongly dated but this time to 1963).

31  Quoted in Suavi Aydın, Kudret Emiroğlu, Ömer Türkoğlu and Ergi D Özsoy, *Küçük Asya'nın Bin Yüzü: Ankara* (Ankara: Dost Kitabevi, 2005), p. 560. The undated photograph of the racing cars ride is reproduced from the BYEGM state archive on p. 559.

32  Uludağ, 'The Social Construction of Meaning in Landscape Architecture', pp. 205–6; Akansel, 'Revealing the Values of a Republican Park', pp. 108–9; Memlük, 'Inclusivity of Public Space', pp. 88–9.

33  Akansel, 'Revealing the Values of a Republican Park', p. 109–10; Memlük, 'Inclusivity of Public Space', pp. 88–9, 94–5. For a photograph of the 'Mehmetçik' miniature train at the station next to the Luna Park see http://www.mimdap.org/?p=28423 accessed 28 September 2014.

34  Uludağ, 'The Social Construction of Meaning in Landscape Architecture', pp. 205, 209; Akansel, 'Revealing the Values of a Republican Park', pp. 108–9; Memlük, 'Inclusivity of Public Space', p. 96 (Figure 4.33). For a present list of rides and photographs see http://malcolmsturkeytrip.blogspot.com.tr/2012/09/ankara-lunapark.html accessed 27 September 2014.

35  For more on this theme see Gürel, 'Architectural Mimicry, Spaces of Modernity', pp. 165–72.

36  Starting with Uludağ, 'The Social Construction of Meaning in Landscape Architecture', pp. 204, 239.

37  Akansel, 'Revealing the Values of a Republican Park', pp. 1–2, 103, 149–50, 154; Memlük, 'Inclusivity of Public Space', pp. 92–3.

38  Ibid., p. 156; Ibid., pp. 57, 66, 92–3.

39  Aydın et al., *Küçük Asya'nın Bin Yüzü: Ankara*, pp. 559–60. The evidence for degeneration comes from a first-hand account. Compare this, however, to the contrasting reminiscences of Buket Uzuner, quoted in the same source (see n. 31).

40  Uludağ, 'The Social Construction of Meaning in Landscape Architecture', p. 182; the fact that the Luna Park was in the first design is also confirmed by Demir, 'Toplumsal Değişme Süreci İçinde Gençlik Parkı', p. 75.

41  Memlük, 'Inclusivity of Public Space', pp. 92–3.

42  See, for example, Burçak Evren, *İstanbul'un Deniz Hamamları ve Plajları* (Istanbul: İnkılâp Kitabevi, 2000) on sea bathing houses and beaches – reviewed by B. Nilgün Öz, *Journal of Tourism History* 1, 2 (2009), pp. 171–3; and an unpublished thesis by Yasemin Keskin, 'İstanbul'da Eğlence Hayatı (1923–1938)' (Master of Arts in Turkish History, Marmara University, Istanbul, 2006) on entertainment in Istanbul in the early Republic.

43  Memlük, 'Inclusivity of Public Space', pp. 2–3, 67, 92–3; Aydın et al., *Küçük Asya'nın Bin Yüzü: Ankara*, p. 559.

44 Uludağ, 'The Social Construction of Meaning in Landscape Architecture', pp. 228–9; Memlük, 'Inclusivity of Public Space', p. 106.
45 The only other threat to the Luna Park at this time was a fire in August 1989 in the 'fear tunnel' (ghost train) which also destroyed adjacent pavilions: Uludağ, 'The Social Construction of Meaning in Landscape Architecture', p. 238.
46 For a particularly judgemental and melancholic portrait of the park at this time see Güven Dinçer, *Ankara: Kent Yazıları* (Ankara: Ankara Üniversitesi Yayınları, 2006), pp. 277–81.
47 Güven Arif Sargın, *Atatürk Kültür Merkezi Değerlendirme Raporu 21.09.2005* (Ankara: TMMOB The Chamber of Architects Ankara Office, 2005), pp. 8–9.
48 Akansel, 'Revealing the Values of a Republican Park', pp. 126, 144, 149.
49 Ibid., pp. 149–50; Memlük, 'Inclusivity of Public Space', p. 69.
50 Memlük, 'Inclusivity of Public Space', pp. 126, 129–30, 135. See also http://www.mimdap.org/?p=28423 accessed 1 October 2014.
51 Batuman, 'Early Republican Ankara', p. 662. The article offers a critical analysis of contemporary political meanings of Ankara as an object of urban historiography and nostalgic yearning.
52 Epitomised in works like Bernard Lewis, *The Emergence of Modern Turkey*, 3rd edn (Oxford University Press, 2002) which has established itself, since first published in 1961, as the classic text on the history of the late Ottoman and early Republican eras.
53 Sibel Bozdoğan and Reşat Kasaba (eds), *Rethinking Modernity and National Identity in Turkey* (Seattle: University of Washington Press, 1997), p. 5. To be fair, the author of the first key reference on the Youth Park does apply this kind of thinking in a later work and related context: Zeynep Sökmen Uludağ, 'The Evolution of Popular Culture and Transformation of the Urban Landscape of Ankara', *TRANS. Internet-Zeitschrift für Kulturwissenschaften* 15 (2003): http://www.inst.at/trans/15Nr/01_2/uludag15.htm accessed 2 October 2014.

# 6 Knott's Berry Farm

## The improbable amusement park in the shadow of Disneyland

*Gary Cross*

When many consider the American amusement park, they conjure images of Coney Island around 1900; others might think of the highly themed Disneyland of 1955, or even of one of the newer corporate amusement parks like Six Flags dating from the 1960s and 70s. Yet, since the 1970s, one of the consistently top ten parks in the United States has been Knott's Berry Farm, the curious emanation from a Depression-era fruit stand and chicken restaurant located on a country road in rural Orange County, twenty-five miles east of Los Angeles, which somehow morphed into a major amusement park.

The contrast with Disneyland is inevitable. Unlike Walt Disney, Walter Knott (1889–1981) was a local farmer, had no experience in movies, cartoon making or even interest in creating an alternative theme-based amusement park. Instead, in 1940, he built a Ghost Town, a tribute to the 'Old West', as a free attraction for waiting restaurant customers. In competition with nearby Disneyland (in Anaheim, a mere seven miles south on the freeway), Knott and his children very gradually added other attractions; first extensions of the Ghost Town, and then, after enclosing the Farm in 1968 and charging admission, an array of themed zones and thrill rides. Knott's youngest daughter assumed leadership in the 1970s and built two themed areas – the Roaring 20s and Fiesta Village – but what attracted crowds were the new steel rollercoasters. Although sold to Cedar Fair in 1997, Knott's Berry Farm had long before been fully transformed into a major thrill ride park.

While many traditional parks evolved from late nineteenth- and early twentieth-century picnic parks (for example, Cedar Point in Ohio[1] and Hershey Park and Knobel's in Pennsylvania), and from the late 1960s corporate sites were heavily capitalised and more or less complete from the start, Knott's Berry Farm had the distinction of emerging out of a roadside attraction or 'tourist trap' designed to draw cars loaded with excursionists.[2] Although the Farm adapted Disney theming and thrill ride innovations from the 1960s, obliterating Walter Knott's original formula, all the Western theming remains; a fascinating composite of its long and varied history. Disneyland is often seen as the template of the modern American theme/amusement park, but Knott's Berry Farm is more typical, more 'accidental', less linked to the film industry and, in a living contradiction to early 'uplifting' historical intents, gradually became a centre of ultra-modern thrill ride technology. This chapter will explore this curious and incongruous history and in conclusion offer brief comparisons with other American amusement parks that emerged about the same time.

## Walter Knott and his world

Walter Knott is certainly an unlikely founder of a major amusement park. Born in 1889 in San Bernardino California to a family of ranchers and prospectors (his grandparents arrived as pioneers from Texas after the Civil War), Knott was brought up by a grandmother in Pomona (after his father died when he was six) who introduced him to the lore of the 'Wild West'. Left without property, Knott rented land as a teenager for commercial gardening, though at the age of twenty he also worked as a bookkeeper. In 1911, after building a house, he married Cordelia (1890–1974) but immediately traded the house for 160 acres of land on the Mojave Desert in hopes of success as a farmer. After about three years, he realised the folly of his arid homestead and worked as a sharecropper and took day jobs (including working at the Calico mine in California in 1917, a site he would later restore). In an effort to save money in order to purchase good farm land, the Knotts went to the Sacramento valley in northern California for three years before returning to Orange County to lease 100 acres to truck farm in 1920. Like others, he sold his produce, especially berries, from a roadside stand near the village of Buena Vista on Highway 39, the main road between Los Angeles and the beaches of Orange County. By 1927, he had finally saved enough to buy ten acres for berry gardening, serving berries and cream as well as slices of berry pie from a five table 'tea room' to pay off his loan. Cordelia added fried chicken and biscuit dinners in 1934 (served at first, according to oft-told Knott accounts, on the couple's wedding china).

Knott was fortunate enough to encounter Rudy Boysen (superintendent of parks in Anaheim) who had created a hybrid berry plant in 1927 that produced an especially large and later ripening fruit, soon known as the Boysenberry. Unable to patent his discovery, in 1932 Boysen gave Knott some wilted plants, which the ambitious farmer nurtured into the centrepiece of his berry and fruit farm. By 1939, Knott had expanded his farm to 100 acres and his restaurant sat 600; adding a gift shop and market store to sell jams and local fresh fruit and vegetables. Contributing to the roadside appeal was a miniature model of a volcano, wishing well and replica of George Washington's Mount Vernon fireplace. This chronicle of persistence and hard work shaped Knott's image and identity; it intrigued journalists from the late 1930s, who, doubtless with Knott's active cooperation, touted the Knott's Berry Place (as it was then known) as the quintessential site of western hospitality, wholesomeness and family enterprise where Knott tended his gardens, his wife cooked (eventually with the help of hundreds of 'neighbours') and his son and three daughters worked at the market place, restaurant and eventually the roadside attractions. In a 1952 magazine interview, Knott boasted of his Depression-era achievement that 'Anyone could do what I've done' – with honest labour and a loyal and dedicated family. It is this story of rugged rural individualism that set Knott's Farm apart from the corporate park developers and offered a distinct variation of Disney's cult of personal achievement.[3]

Knott's enterprise and 'pioneer spirit' continued in the 1940s when he transformed two acres adjacent to his restaurant and store into a Ghost Town, a curious blend of authentic buildings transported from abandoned south-western boom towns and typical tourist trap whimsy (and, in this case, racism). As a place for people to go while waiting for seats in the ever-crowded restaurant, Knott first offered a Covered Wagon train cyclorama (a miniature set of wagons set in a painting) located in the Old Trails Hotel, built in 1868 and transplanted from Prescott Arizona to the Farm. For what Knott called 'a moment of reverence' in a 'quiet farm atmosphere', he also built the Little Chapel by the Lake with a featured painting by a local German-born artist, Paul von Klieben, *The Transfiguration of Christ*.[4] Von Klieben was also architect of most of the early Ghost Town that expanded to three streets on which were erected the obligatory saloon, news office, fire department, hangman's tree and jail (Figure 6.1). Western stereotype figures were central features (including a lifelike Wing Lee in his Chinese laundry and Sad Eye Joe in jail, and even an Indian playing poker while holding five aces). In 1948, Knott constructed a miniature Gold Mine with stamp mill, sluice way and fake miner that beckoned children with the offer to pan for 'gold' (Figure 6.2). Later, in 1954, came a ticketed attraction, the Haunted Shack with the classic 'fun house' trick of water running up hill. Other attractions included a narrow gauge railway (1951) that circled the site (moved from Colorado), stage coach rides, the Bird Cage Theater (for live shows re-enacting corny nineteenth-century melodramas where the audience would boo the villains and cheer the heroes), Old MacDonald's Farm (a petting zoo for kids) and the Overland Trail Ride (a horse-pulled surrey through a 'pioneer trail' with 'reminders of bad water, Indian raids, sick animals and fires'). Knott even put his original Berry Stand and second-hand Model-T Ford that he drove in the 1920s onto the site (Figure 6.3).[5]

But Ghost Town also included buildings where traditional crafts were displayed (glass blowing, blacksmithing and so on), as well as displays of late nineteenth-century dry goods in the General Merchandise store with its pot-bellied stove and cracker barrel, along with the rather extensive collection of photographs and tools in the Western Trials Museum.[6] All this was typical of the authenticity sought by contemporary heritage sites like Colonial Williamsburg or Sturbridge Village.[7] Knott took his Ghost Town seriously. The cyclorama was to be 'a tribute to my family and their coming across the plains in covered wagons' and the broader project was 'to show how little those people had to work with and how much they had been able to accomplish'.[8] While Disney promised that his Main Street USA (Middle America circa 1900) would take elders back to their youth and inspire their grandchildren, Knott offered a similar vision of nostalgia for the Wild West.[9]

Still, Ghost Town was more a roadside novelty than a heritage site. Knott knew that families were looking for fantasy and whimsy, not a history lesson. Across the road from Ghost Town, Knott had Forest Morrow erect a kid-friendly attraction, Jungle Island, with 140 animal-like figures (woodkins) made from logs and

*Figure 6.1* Knott's Berry Farm's Ghost Town

Source: © The editor's collection

*Figure 6.2* The Ghost Town miniature Gold Mine

Source: © The editor's collection

*Figure 6.3* Walter and Cordelia Knott in front of their original Berry Stand and beside their old Model-T Ford

Source: Knott's Berry Farm postcard © The editor's collection

branches.[10] Knott did, however, follow through on his goal of preserving the old West by undertaking the restoration of Calico (the site of his uncle's prospecting efforts in the 1880s), ten miles east of Barstow, which in 1966 he turned over to county authorities as a park.[11] But Knott kept down the didactic tone of his own Ghost Town. That was for profit.

The 'adventure' of Ghost Town was not enough by the end of the 1950s. Although Knott claimed that he was not at all threatened by the arrival of nearby Disneyland in 1955, he commissioned a survey of visitors and found a demand for more entertainment. He reluctantly but immediately bought a vintage carousel (hardly part of the Ghost Town theme) and in 1966 even acquired cable cars (from San Francisco).[12]

Part of the shift toward a wider view of amusement was advanced by the entry of Bud Hurlbut into Knott's inner circle. Hurlbut, along with his sometimes associate Christopher Merritt, were part of a small group of independent ride and attraction builders working at various amusement and 'kiddie' parks.[13] In 1960, Hurlbut talked Knott into removing a horse show to build the Calico Mine ride comprising a rail journey in ore-cars through a six-storey building that Hurlbut co-owned with Knott. The 'dark ride' went through a 'realistic' mountain with portrayals of collapsing beams and explosions (to simulate the dangers of underground mines) but ended in a 'glory hole' full of animated figures (Figure 6.4). After this, Hurlbut planned a Jungle Cruise (similar to the Disneyland ride in Adventureland) but settled on the Overland Trail Ride around Morrow's Jungle Island. Beginning in 1963, Hurlbut began what became his most important project, the log ride. Though starting as a themed coaster, it evolved a few years later into an innovative Water Flume ride: riders seated in simulated fibreglass logs on wheels travelled through a 2,100 foot-long log camp that ended in a 40 foot slide (Figure 6.5). Efficiently, it accommodated 1,600 riders per hour. Opened in July 1969, with the legendary John Wayne and his youngest son riding in the first 'log' followed by festive demonstrations of log rolling and pole climbing, the log flume quickly became a featured attraction at the Farm.[14]

Yet Knott, already 70 years old in 1960, shifted away from interest in the emerging amusement park to focus on his pet project, a replica of Independence Hall that opened in 1967. Built across the road from the restaurant and Ghost Town, this miniaturised reproduction was clearly intended to edify: tours were received in Heritage Hall from where attractive young women in colonial garb led groups to a theatre for a patriotic film, with a strong emphasis upon limited government, and thereafter into Independence Hall. There visitors heard live debates about the Declaration of Independence and saw an original copy of the Declaration, a prized Knott possession. Knott always insisted that the Farm was more than an amusement park.[15]

During the late 1960s and 70s, Knott, already a conservative Republican, became very active in the Orange County right. Hostile to government, he refused Social Security payments, vigorously opposed unions and complained about the high marginal tax rate impeding economic incentive. He promoted the California

*Figure 6.4* The Calico Mine

Source: Knott's Berry Farm postcard, Bob Ellis © The editor's collection

*Figure 6.5* The Water Flume and Ghost Town's train and stagecoach in Calico Square

Source: Mitock & Sons postcard, Mike Roberts © The editor's collection

Free Enterprise Association and invited Presidential candidate Richard Nixon to the Farm in 1960. His conservatism was more than political: the *Los Angeles Times* claimed that he still wore high button shoes in 1969, continued to live in the small adobe house he built in the 1920s and that his office was adorned in American flags, Republican elephants and busts of Abraham Lincoln. But, by then, Knott had ceded control to his children.[16]

## New leadership

First, the son Russell, and then the youngest daughter Marion, gradually led the family enterprise. Born in 1925, and having worked at the Farm continuously since she was ten, graduating to store management and merchandising in her twenties, Marion was the most influential of the children in broadening the scope of the Farm in the 1960s. Their most momentous move was to enclose the park and to charge an entrance fee (initially 25 cents) in 1968 (with a Bonanza Fun Book costing another three dollars for admittance to six rides and attractions within).

This was in line with the practice of Disneyland and other parks. The street carnival model of pay-as-you-go, with separate admissions to each attraction, inevitably attracted undesirable youth, often 'free loaders' who milled in the crowd but spent little and even more drove away middle-class families with small children and money to spend. The presence of these youth (especially minorities) was one of the key reasons for the downfall of the amusement parks at Coney Island in the 1940s and 60s. Disney had avoided this problem when he built his park in Anaheim, creating an ideal consuming crowd, consisting primarily of indulgent and nostalgic parents and desiring kids, with his thematic Disney attractions but also with park enclosure and (soon after opening) the introduction of a fixed entry ticket.[17]

Like other new style park developers Marion adopted this model. As she explained in the house publication *Knotty Post* in 1968, the object was to replace the 'boisterous crowds of youngsters with their blaring transistor radios with just nice adults and nice kids having fun the way fun should be had'.[18] This comment certainly reflects the hostility of a middle-aged woman to the pop culture of teenagers in the late 1960s, and possibly fear of minorities who appeared as the stereotypical carriers of portable radios playing loud music at the time, but it also reflects how she had reconceived of Knott's Berry Farm: no longer was it a chicken restaurant and specialty food store with an unusually large roadside attraction.

Marion was also intent on breaking from the Wild West and patriotic themes of her father. Not only was the Covered Wagon Camp dismantled in 1968 but the first major innovation as the Farm transitioned into a modern middle-class amusement park was the construction of Fiesta Village, which opened in June 1969. Replacing Old MacDonald's Farm, the Village began as merely a square of shops and eateries dressed in red tile roofs surrounding a 'Lake of Flowers' and Aztec fountain, enlivened by strolling musicians and nightly concerts by a marimba band. It was to 'capture the gaiety and grandeur of fiesta time, all the time'. Gradually, themed rides for small children were added to the mix: a Fiesta Wheel, Happy Sombrero (a circular ride similar to Disney's Dumbo ride) and the Mexican Whip (a themed version of a common carnival ride).[19]

The Fiesta Village was still a celebration of Old California with its Mexican heritage but Marion wanted to go further by developing a zone deeper into the park that eventually was called the Roaring 20s. Typically, this was not the first plan. In 1972, the Farm hired Rolly Crump (who had formerly worked for Disney and the aborted theme park Circus World in Orlando) to design a gypsy camp show-ing how gypsies lived. This encampment was to be set on a hill and surrounded by a 'cave of arcade games, magic shops, and a gypsy orchestra on Weeping Rock stage'.[20] This plan was soon transformed into a dark ride where the gypsies became comical bears that Marion wanted to be dressed up as cowboys (following her father's Western theme) but in the end they became bear gypsies (mechani-cal figures viewed from cars) in a children's ride called Bear-y Tales with jumpy theme music. This ride became an odd attraction at the edge of a new project, the Roaring 20s.[21] The central building was the Funky Fabulous Airfield Eatery, a café set in an authentic airplane hangar from the days of the biplane. Setting the mood were Dixieland bands and even a flagpole sitter for the publicity-filled opening in July 1975. In a 1976 press release, Marion touted this zone as a miniature of 'an amusement park which could well have existed in California during the 1920s' to inspire the old seeking a recollection of their 'golden years' and for the 'younger caught up in the nostalgia of that era'. She added that the Roaring 20s was a 'trib-ute to her parents' just as her father's Ghost Town had been a commemoration of his parents' and grandparents' lives as settlers in California.[22] Yet, from the beginning in 1975, central features of the Roaring 20s area were the Corkscrew coaster (featuring a 70-foot drop to a double loop) (Figure 6.6) and Whirlwind (a sweeping circle coaster imported from France), as well as classic rides like bumper cars. Marion explained that it was the Knott's 'duty to keep up with the times and to consistently offer the newest and finest in family enjoyment'; and that the Corkscrew was more than a state-of-art coaster for those seeking the thrill but appealed to those 'also looking for something with history and nostalgia to it', set as it was in the Roaring 20s 'airfield'. It was to be a place 'where three genera-tions can spend a whole day together'.[23]

Yet, throughout the 1970s, as elsewhere, there was an obvious shift toward the thrill ride. This proved highly successful for Knott's Berry Farm. In its inaugural year, the Corkscrew had 2 million riders as the Farm began to attract a teenage and youth audience.[24] Still, the Knott family insisted that it remained rooted in tradition. The 1976 press release for the spring crowds continued to stress the 'old time adventures' of the gold panning attraction, the daily re-enactments of an old West 'holdup' and the fun of riding the Happy Sombrero at Fiesta Village. Publicity continued to tout the loyalty of Farm employees like Hattie Bilbrey who at 85 in 1979 continued to demonstrate the art of spinning in Ghost Town as she had for decades. Even in the late 1970s chickens and ducks roamed the park freely and the Farm still had 40 stagecoach horses and 20 burros for its throwback rides (something that Disney had long abandoned as inefficient and too costly). It is revealing, however, that Farm manager Guy Tester admitted that 'country' themes no longer came up in staff meetings. But the site still had to uphold its 'brand'. Walter Knott's obituary in 1981 stressed his pride in the Farm's repro-duction of Independence Hall[25] and as late as 1984, a third of the Farm's revenue still came from the chicken restaurant and other food outlets.[26]

And like other parks, Knott's Berry Farm continued a tradition of live entertain-ment, especially of music and other attractions that appealed to a slower-paced,

*Figure 6.6* The Corkscrew coaster in the Roaring 20s area. The first thrill ride of its kind, the Corkscrew carries passengers up a 70-foot incline, then twirls them completely upside-down twice

Source: Knott's Berry Farm postcard, Mike Roberts © The editor's collection

older conservative audience. The keystone of this policy was the John Wayne Theater that opened in 1971. With a lobby that featured artefacts and movie stills from John Wayne's many cowboy movies, the auditorium provided 2,150 seats and modern lighting and sound systems for live shows like the Country Western Shindig held in February 1972.[27] In the 1970s, the Farm reached out to a somewhat younger crowd with a series of 'Halloween Haunts' programmes featuring famed disc jockey Wolfman Jack and his Shock and Rock Revue in 1975.[28] Picking up on the fad for citizen band radios, the Farm presented the Van and CB Happening in March 1977 with country music, CB celebrity C. W. McCall and displays of brightly decorated vans (popular at the time), along with Highway Patrol officers who lectured the crowd on the legal uses of CBs for police work.[29] While Disneyland in the 1960s and 70s featured concerts of swing and Dixieland music for middle-class elders nostalgic for their youth in the 1930s and 40s,[30] Knott's sought a niche with conservative rural America.

However, efforts to capitalise on the rising tide of the thrill ride while continuing to appeal to Walter Knott's traditional clientele grew increasingly difficult. Not only were the traditionalists dying off but appealing to a rather narrowly conservative sector of the population severely limited the Farm's potential customer base and defeated its outreach to the thrill generation. While Knott's drew a gate of more than 3 million in the mid-1970s, making it the third most popular amusement park in the United States, by 1979 (with only 2.9 million customers) the Farm had dropped to sixth place behind Six Flags/Great America near Chicago and Busch Gardens–Tampa, as well as the Disney and Universal Studio parks in Orlando and Anaheim.

Not long after the death of Walter Knott in 1981, Terry van Gordon emerged as leader of development at the Farm. He orchestrated a 100 million dollar renewal and expansion in the 1980s that was followed by even more radical changes in the 1990s, transforming the Farm into a fully modern amusement park that reached a wide audience.[31] The new management followed a multi-pronged attack: one the first moves was to develop a children's zone. In 1982, van Gordon tore down the outmoded Jungle Island playground outside the park gates (where there were no rides, just exploring and playing games) to make room for administrative offices. But he also signed an agreement with Charles Schultz for 'Camp Snoopy', a collection of rides and live attractions for under 12-year-olds. Since 1983, this has been a model of the modern kiddie zone in amusement parks, designed for the small fry but whimsical and not at all traditional. For example, in 1996 'Snoopy's Most Excellent Magical Adventure' was performed daily at the Knott Toyota Good Time Theater.[32]

Knott's was hardly laggard in joining the fad for 3D movies, introducing Sea Dream in 1987, a 23-minute undersea fantasy at the Cloud-9 Ballroom Theater in the Roaring 20s area. And in 1988, the park introduced a themed water ride, Big Foot Rapids, a 10 million dollar attraction on 3.5 acres featuring a six-minute trip on 'rafts' during which riders 'hunt' for Big Foot. This attraction became the anchor for the Wild Wilderness section of the park. And in 1992, the Farm radically updated the Western theme of Ghost Town with the two acre Indian Trails 'celebrating the arts, crafts and traditions of native Americans'. Featured displays included artefacts from Plains and Far Western tribes (tipis and totem poles), representing van Gordon's attempt to portray 'the creativity of native American culture with dignity and respect', in marked contrast to the neglect and carica-tures of native peoples in Ghost Town. A novel addition came with the Mystery Lodge, a multimedia spectacular, centred on the hologram figure of an 'Old Story Teller'.[33] The 1993, Kingdom of the Dinosaurs was part of another modern trend toward the quasi-realistic dark ride: riders travelled in a 1928 Los Angeles trolley car through the cold of the ice age, heat of volcanoes and back in time to the age of dinosaurs. The press release for this ride boldly claimed that 'Knott's Berry Farm combines entertainment and authenticity for an exciting adventure in education. Designers research in museums and interview to get it right'. The set designers came from Hollywood studios, of course.[34]

But the key to the revival of the Farm was the expanding programme of thrill rides, especially tubular steel rollercoasters. Building on the decision in 1978 to place a coaster amusingly called Montezooma's Revenge in the once rather sedate surroundings of Fiesta Village, the park added the Jaguar to the Village in 1995, a fully themed ride entered through a 'Mayan pyramid' and experienced with twenty-four others on a jaguar-shaped vehicle 'simulating the sensation of a great jungle hunt' that cuts over the park.[35] Marion's pet project, the Roaring 20s was updated in 1996 as the Boardwalk, a site that 'salutes the vigor, vitality, and vari-ety of southern California's legendary seaside culture' (a generalised nostalgic reference to an upcoming generation who surely no longer had a particular fond-ness for the 1920s).[36]

By 1996, admissions had climbed to 3.5 million (with 230 million dollars in revenue), making the Farm an excellent target for a buyout. The Knott family had mostly retired or withdrawn from the business and several suitors appeared, including Disney Enterprises. The Farm, however, was sold to Cedar Fair, based at the Cedar Point amusement park on Lake Erie and serving northeastern Ohio (Cleveland, Akron and Youngstown). Inevitably, more changes came quickly: additional coasters like the Windjammer Surf Racers – a dual track racing coaster with a spiralling finish (1997), the Boomerang (1999), the GhostRider (1998), Xcelerator (2002), Silver Bullet (2004) and Sierra Sidewinder – another spinning coaster (2007). In addition, in 2006, the inevitable Knott's Soak City water park opened following a near universal trend at amusement parks.

### Discussion

Knott's Berry Farm today is a fully fledged thrill park superimposed on a rela-tively old roadside attraction and chicken restaurant. The Ghost Town remains but without the free range chickens and ducks and without the dust. It has the look and feel of a theme park fantasy of a Ghost Town – super-clean, with asphalt, fibreglass and, in my visit (admittedly on a slow day in February) with only a few dozen visitors. The museum of pioneer artefacts remains but with even fewer visi-tors and the open-air show (complete with comical cowboy shootout and country music) has three gigantic space-age orange tubes and white pillars of one of the rollercoasters as a backdrop. The themed zones seem like randomly placed stage sets between which the coasters weave. Independence Hall survives but behind the park, and the restaurant and market remain outside the park.

In the eyes of most, Knott's Berry Farm is an anomaly. It is not a survivor (like Kennywood in Pittsburgh) from the era of the streetcar/subway parks. The Farm lacks their ties to Coney Island (Steeplechase Park of 1897, the much larger themed Luna Park of 1903 and the even more elaborate Dreamland of 1904) and their many imitators from Chicago to Butte Montana.[37] Nor is Knott's Berry Farm part of the corporate parks constructed in the 1960s and 70s, much less a planned park heavily themed by film studios (Disneyland, Walt Disney World and Universal Studios). Yet, in many ways, it paralleled the development of both the corporate parks and the Disney/Universal Studio phenomenon. The earliest of the corporate parks was Six Flags over Texas, opened in 1961 in Arlington, Texas, and followed by Six Flags parks in Atlanta (1968) and St Louis (1971). Though themed around different periods of American history (represented by the six flags), these parks were especially noted for their new style coasters, abandoning the wooden tracks and frame of the traditional 'woodies' for tubular steel tracks (a technology ironically perhaps introduced to the United States in the heavily themed and relatively sedate 'Matterhorn' attraction at Disneyland in 1959). Others followed like Magic Mountain in Valencia California (1971) and King's Island in Cincinnati (1972). In 1976, J. Willard Marriott Sr of hotel fame opened the Great America amusement park on a 200 acre farm in Gurnee, Illinois, north of Chicago. Although themed with a simulated Yankee fishing village, Yukon mining camp and a New Orleans French Quarter, what gathered the crowds were its coasters and water flumes. By 1980, Great America was attracting about 3 million visitors in a five-month season. What made these parks unique was not only their well-financed construction (in contrast to older family parks) but their appeal to a new generation of youth for whom the thrill rides appealed. This separated these new parks from the family-friendly environs of Disney and other child-themed parks with lots of 'kiddie rides' (Idlewild near Pittsburg and, of course, later Legoland at San Diego in 1999).[38] Yet, even under family ownership, Knott's Berry Farm kept up with this trend by introducing the cutting-edge Corkscrew with its spiral loop in 1975 and continued to do so thereafter, long before Cedar Fair's takeover.[39]

Obviously, the Farm contrasted with the sophisticated theming of rides at Disneyland. The Knott family made it clear that it was traditional and held on to its down-home, local image compared with the global reach of Disney, first in Anaheim, then in Orlando, Paris, Tokyo and, eventually, Hong Kong. And the admission price was always much less.[40] But in the 1980s Knott's Berry Farm followed Disney in trying to broaden its appeal beyond its 1950s image. Disney, pressured by the exciting rides offered at the relatively new corporate parks (like Six Flags) and modernised traditional parks (Kennywood and Cedar Point), but even more by Universal Studios (especially in Orlando in the late 1980s), slowly accommodated the trend toward thrill rides. These included The Tower of

Terror in 1994 and indoor Rock 'n' Roller Coaster in 1999 in its Orlando parks, even though Disney still tried to disguise thrill rides as part of a story.[41] In 2006, even the innovative environmental zoo of Disney's Animal Kingdom introduced a themed coaster, Expedition Everest, with a none-too-persuasive backstory about riders pursuing a yeti.[42] Most amusement parks went for the thrill rides as youth demanded more than the themed attractions and tame rides of the original Disneyland and Knott's Berry Farm.

In this transformation, Knott's Berry Farm had a lot in common with other parks that emerged after 1968 (when, with its enclosure, the Farm became a modern amusement park). Consider Magic Mountain that opened in 1971. Built by George Millay of Sea World in Valencia, California, it was at first a very traditional family amusement park with a carousel dating to 1912, conventional coaster, log flume and Victorian-era gazebo where a Dixieland band played. Later, the park featured oldie groups for the nostalgic (for example, the Mills Brothers and Sha Na Na in the summer of 1976).[43] But also, in 1976, Magic Mountain introduced The Great American Revolution, and two years later the Colossus – two high, fast, looping steel tube coasters like Knott's Berry Farm's Corkscrew. In 1979, Magic Mountain was sold to Six Flags and became part of the thrill park culture that dominates today. Knott's Berry Farm may have held on to its Ghost Town, Fiesta Village and the Roaring 20s (Boardwalk) themes but the place paralleled the thrill trend elsewhere. Even more to the point, the Farm was hardly alone in making a dramatic shift from 'family' and traditional fun to thrills in the mid to late 1970s.

Starting out as a roadside berry stand and chicken restaurant in the 1930s, expanding into a typical tourist trap with a maudlin Ghost Town in the 1940s, and rather haphazardly morphing into an amusement park in the late 1950s and 60s, today's Knott's Berry Farm with all of its high-tech coasters could easily be called improbable if not an accident. Yet, in its shift from Walter Knott's celebration of the rustic West and conservative values of his upbringing to an up-to-date amusement park, where the steel rollercoasters draw the crowds rather than the Western Trails Museum, Knott's Berry Farm parallels the transformation of other modern amusement parks. Thus perhaps it, rather than its neighbour, Disneyland is really the more typical.

## Notes

1 David Francis, *Cedar Point* (Charleston: Arcadia, 2004), pp. 20–6.
2 Typical along western American highways at this time were displays of dinosaurs, Paul Bunyan and other storybook characters, and pioneer and cowboy artifacts. For this theme see John Margolies, *Roadside America: Architectural Relics from a Vanishing Past* (London: Taschen, 2010).
3 Walter Knott, *The Enterprises of Walter Knott Oral History Transcript,* interviewed by Donald J. Schippers, 1963 (Oral History Program, University of California, Los Angeles, 1965), pp. 8–16, 34–6, 44–6, 54–60, 63–5; *Los Angeles Times* (13 January 1943), p. 2 and (28 January 1946), A2, in Orange County Archive, Knott's Berry Farm Archives (hereafter cited as OCA-KBF) Clippings Box; Fundinguniverse.com/company histories/Knott's Berry Farm, OCA-KBF, Box 9; quotation from *Magazine of California* (8 November 1952), p. 5; 'The Story of Knott's Berry Place', a publicity

publication of Knott's Berry Place (1940–43) that included clippings of highly favorable local newspaper and magazine reports (including from the *Saturday Evening Post* (2 May 1942), and even a puff piece about Knott's devotion to hard work and family by famed motivational speaker and promoter Dale Carnegie, in *Los Angeles Evening Herald and Express* (10 February 1943), OCA-KBF, Box 8. Norm Hygaard, *Walter Knott: Twentieth Century Pioneer* (Grand Rapids, MI: Zondervan Press, 1965).

4  Fundinguniverse.com, in OCA-KBF, Box 9; Knott, *The Enterprises of Walter Knott,* p. 32.

5  Knott, *The Enterprises of Walter Knott,* pp. 18–23; clippings from *Popular Mechanics* (May 1950), p. 23; *Los Angeles Times* (16 November 1954), A2; OCA-KB, Clippings 2006 Box; *Orange County Life, Business, and Industry* (1963), p. 24, clipping in OCA-KBF, Box 9.

6  Knott, *The Enterprises of Walter Knott,* pp. 25–8.

7  For this theme see Gary Cross, *Consumed Nostalgia: Memory in an Age of Fast Capitalism* (New York: Columbia University Press, 2015), ch. 7.

8  Knott, *The Enterprises of Walter Knott,* p. 68.

9  *Christian Science Monitor* (16 November 1950), n.p., OCA-KBF, Clippings 2006 Box.

10  Jungle Island, like so many other naïve sites that appealed to children, is the subject of considerable nostalgia on the web. For example, see https://www.facebook.com/pages/Knotts-Jungle-Island/118288511518863 accessed 24 May 2014.

11  Knott, *The Enterprises of Walter Knott,* p. 83; Knott's Berry Farm, *Ghost Town* (Buena Vista, CA: Knott's Berry Farm, 1959).

12  KBF publicity sheet (1995), OCA-KBF, Box 8; Knott, *The Enterprises of Walter Knott,* pp. 81–2.

13  Clippings: *Contra Costa Life* (18 May 1969), n.p.; *Los Angeles Times* (30 August 1969), A1; *E Ticket* (a small ephemeral publication concerning the amusement park industry) 35 (Spring 2001), OCA-KBF, Box 9.

14  *E Ticket* (Spring 2001), OCA-KBF, Box 9; KBF publicity sheet (11 July 1969), Box 9; various press clippings regarding the opening of the log ride: *Long Beach Independent* (30 July 1969); *Santa Ana Register* (30 July 1969), *Anaheim Bulletin,* (12 July 1969), OCA-KBF, Box 9.

15  *San Fernando Valley Citizen-News* (30 January 1967), n.p. in OCA-KBF, 2006 Clippings Box; 'Independence Hall', Knott's Berry Farm publicity sheet (1967), OCA-KBF, Box 8; *Los Angeles Times* (22 February 1970), n.p., OCA-KBF, Box 9.

16  Knott, *The Enterprises of Walter Knott,* pp. 106, 113; *Los Angeles Times* (22 February 1970), n.p., OCA-KBF, Box 9; *Star and Herald of Panama City* (23 January 1970), n.p., OCA-FBF, 2006, Clippings Box.

17  For this theme see Gary S. Cross and John K. Walton, *The Playful Crowd: Pleasure Places in the Twentieth Century* (New York: Columbia University Press, 2005), ch. 2, 3 and 5.

18  Marion Knott, *Knotty Post* (Fall 1968), OCA-KBF, Box 2.

19  Publicity sheet (July 1976), OCA-KBF, Box 8; *California Apparel News* (6 June 1969); *Orange County Register* (25 May 1969), E4; *Buena Park News* (11 June 1969), n.p., OCA-KBF, 2006 Clippings Box.

20  *Buena Park News* (9 February 1972); *Holiday Inn Magazine* (November 1972), OCA-KBF, 2006 Clippings Box.

21  Christopher Merritt, 'Riding the Red Line' (2001), in www.christophermerritt.com copied in OCA-KBF, Box 9.

22  *Western Building Design* (October 1976), pp. 6–7, OCA-KBF, 2006 Clippings Box; KBF publicity sheet (6 July 1975); KBF publicity sheet (5 February 1976) (for interview with Marion Knott), OCA-KBF, Box 8; *Van Nuys, Valley News* (6 July 1976), OCA-KBF, 2006 Clippings Box.

23  *Long Beach Independent* (7 June 1975), 2006 Clippings Box; KBF publicity sheet (5 February 1976), OCA-KBF, Box 8; *Sundancer* (May 1978), n.p., OCA-KBF, Box 9.

24  Clipping: *Sundancer* (May 1978), n.p., OCA-KBF, Box 9.

25  1976 publicity sheet, OCA-KBF, Box 8; *Los Angeles Times* (22 October 1979), OC-A1; *Orange County* Register (12 December 1979), n.p.; *Buena Park News* (5 December 1981), n.p.; *Orange County Register* (3 May 1977), n.p.; OCA-KBF, 2006 Clippings Box.

26  Clipping: *Los Angeles Times* (13 May 1984), HI, OCA-KBF Box 9.

27  Variety (8 June 1971), n.p.; *Box Office* (30 August 1971), pp. 28–31; *Buena Park News* (9 February 1972), OCA-KBF, 2006 Clippings Box.

28  KFB publicity sheet (28 October 1975), OCA-KBF, Box 8.

29  KFB publicity sheet (5 March 1977), OCA-KBF, Box 8.

30  From the Anaheim Public Library, Disney Resort Room: 'Disneyland Diary: Year Five' (1973), Disneyland 1960 file: 'Special Events', *News from Disneyland* (1965); 'Big Bands', *News from Disneyland* (May 1967), 'Performers at Disneyland' and 'Senior Citizen Days', *News from Disneyland* (Spring 1974).

31  Fundinguniverse.com in OCA-KBF, Box 9.

32  Publicity sheet (1996), OCA-KBF, Box 8.

33  KBF publicity sheets (1987, 1988, 1992, 1993, 1994), OCA-KBF, Box 8; Christopher Merritt posting (12 April 2004), copy in OCA-KBF, Box 9. See David Kamper, 'Cowboys and Native Americans: Locating Culture, Identity and Agency at Knott's Berry Farm' (unpublished MA thesis, UCLA, 1999) for a critical assessment of the Mystery Lodge and Indian Trails.

34  KBF publicity sheet (1993), OCA-KBF, Box 8.

35  KBF publicity sheet (1994), OCA-KBF, Box 8.

36  KBF publicity sheet (1996) OCA-KBF, Box 8.

37  Gary Kyriazi, *The Great American Amusement Parks: A Pictorial History* (Secaucus, New Jersey: Citadel Press, 1976), pp. 47–57; Oliver Pilat and Jo Ranson, *Sodom by the Sea, An Affectionate History of Coney Island* (Garden City, New York: Doubleday, 1941); pp. 144–6; Woody Register, *The Kid of Coney Island: Fred Thompson and the Rise of American Amusements* (New York: Oxford University Press, 2001), pp. 92, 132–3.

38  Scott Rutherford, *The American Roller Coaster* (Osceola, Wisconsin: MBI Publishing, 2000), pp. 102–03; 'America's theme parks', *Newsweek* (4 August 1980), p. 56; 'Roller-coasters. Hold on to your hat', *The Economist* (24 February 1996), p. 87.

39  'Knott's Berry Farm upgrades with new thrills and themes', *Architectural Record* 187, 11 (November 1999), p. 50.

40  In 2014, adult (over 10 years old) admission to Disneyland was 96 dollars compared with the 65 dollars (with seasonal discounts) of Knott's Berry Farm.

41  'New attractions in honor of Mickey', *USA Today* (1 June 1988), 4d; 'Rock 'n Roller Coaster', *Eyes and Ears* (Walt Disney World in-house newsletter) (23 April 1998), pp. 1, 3.

42  'Body Wars', *Disney News* (Spring 1989), p. 36; 'Rock 'n Roller Coaster', *Eyes and Ears* (23 April 1998), pp. 1, 3; 'On Track', *Disney Magazine* (Fall 1998), pp. 44–7.

43  'Sea World and Magic Mountain, 1971–1976', George Millay Papers, University of Central Florida Archives.

# 7 The dilemma of the crowd

## Atlantic City's Steel Pier, George Hamid, and leisure and urban space in post-civil rights America

*Bryant Simon*

'The Jersey coast', the travel writer Harrison Rhodes wrote after a visit to Atlantic City in 1915, 'is the most popular part of the American seashore, the most characteristic, the most democratic, the most intensely American'.[1] The British author recognised that Atlantic City thrived because it reflected the American mainstream. Like all mass resorts then and now, Atlantic City, in its good days when it swelled as the 'Nation's Playground' and at its lower point when you could roll a bowling ball down the Boardwalk and not hit a soul, mirrored the dreams and anxieties of the vast American middle classes. Turning these hopes and fears into bricks and mortar, Ferris wheels and dining rooms was the job of the city's tourist entrepreneurs. What they did, whether they acknowledged it or not, was to try to give form to people's private thoughts and desires; the best of them like Coney Island's Fred Thompson and California's Walt Disney, sometimes understood leisure consumers better than they understood themselves.

If the people in the business of leisure wanted to survive, however, they had to know when the private thoughts of their patrons changed and they had to change their amusement parks and destinations with them. Change, though, was never easy; not for the people in the stands or for the tourist entrepreneurs trying to get them to spend their money. But that is why studying the Fred Thompsons and Walt Disneys of the world is worthwhile, their triumphs and flops reveal much about them and the people they targeted to come to their fun places and fantasy spots.[2]

George A. Hamid (1896–1971) was Atlantic City's Thompson and Disney, and his Steel Pier was a smaller, though only slightly smaller, version of Luna Park and Disneyland. In the early days of the mid-twentieth century, when the Boardwalk was so crowded that it resembled a subway platform at rush hour, the amusements stretched, in one observer's exaggerated description, from 'the New Jersey shoreline [to] somewhere near the coast of Spain'.[3] While Steel Pier was not, of course, that big, it was a mass amusement site, buzzing with lights and activities and throngs of well-dressed couples and their kids. It had in those days something for everyone, when something for everyone worked, and drew large crowds to America's leisure sites (Figures 7.1 and 7.2).

*Figure 7.1* The Boardwalk and Steel Pier

Source: Tichnor Brothers Inc. postcard © The editor's collection

*Figure 7.2* Large crowds mob the Boardwalk and the front of Steel Pier, 1935

Source: Fred Hess & Sons © ACFPL Atlantic City Heritage Collection

Towards the front of Hamid's Steel Pier there might be incubator babies and a giant Goodyear tyre on display. Off to the side, General Motors operated a showroom for its latest models and most popular cars. Next to the Chevys and Oldsmobiles were boxing cats and racing goldfish (Figure 7.3). Moving towards the middle of the Pier, visitors found the picnic area where they left their coolers and boxed lunches. (Atlantic City laws forbid the sale of food on the piers.) Also towards the middle were the movie theatres where three 'first run selected motion pictures [ran] each day'. All of them, George Hamid assured customers, 'are chosen by a special board of review with the purpose of providing the finest entertainment for all members of the family'.[4] Further away from the Boardwalk, Hamid and his crew presented minstrel shows and vaudeville acts, comics, magicians, sword swallowers and pole sitters. At the back of the Pier, hanging out over the ocean, was the gold-domed ballroom where the biggest of the bid bands – Glenn Miller, Harry James and Benny Goodman – played. When they finished, child stars wowed the crowds. At the very end of the Pier was the 2,500-seat Water Circus amphitheatre, where following water-skiing dogs and human cannonballs the celebrated diving horse performed (Figures 7.4 and 7.5). Five times a day during the week and seven times on the weekends, a woman in circus sequins plunged atop of horse into a small pool of water. Customers got all of this and more, for one relatively low price.

By the late 1960s, however, George Hamid had a problem on his hands. Steel Pier still stretched far into the ocean. It still had a ballroom at the end, and it still employed women on diving horses. But what it did not have were the crowds; crowds big enough to keep the business viable. So, Hamid had to work out how to reconfigure his amusement pier and find new attractions to either bring back the old crowd, or build a new one, or to create some combination of the two. How he thought about his business problem reveals much about the economics of leisure but also about race and space in modern America, because a man like George Hamid did not just sell admission tickets, he sold something much more elusive – private dreams and fears. When he could not tap into those emotions, he could not make money.

## George Hamid

George A. Hamid Sr billed himself as the 'man who worked himself [up] from the bottom to the top'.[5] Born in a tiny Lebanese village, the owner of the Steel Pier started out performing in the streets, earning coins to help pay for food for his family. Broad shouldered and remarkably strong and powerful for his age, Hamid joined his uncle's high-flying circus act when he was only nine years old. Sometimes, he anchored human pyramids; other times he flew through the air with the trapeze artists. The legendary showman (and himself a tourist entrepreneur) Buffalo Bill Cody saw the young Hamid and his uncle's act in the south of France and brought them back to the United States to perform as 'Abou Hamid's Arabs'.

*Figure 7.3* Boxing cats, 1925. This novelty had been a feature of Steel Pier long before Hamid's day

Source: Fred Hess & Sons © ACFPL Atlantic City Heritage Collection

*Figure 7.4*  Night scene at the Water Circus amphitheatre where acts included diving and a human cannonball, early 1930s

Source: Saltzburg's Merchandise Co. postcard © The editor's collection

*Figure 7.5* The diving horse taking off, 1945 – the year Hamid acquired Steel Pier

Source: © ACFPL Atlantic City Heritage Collection

Buffalo Bill did more than hire the young Lebanese performer. He kept him close by, often referring to him as an adopted son even though he treated him more like a servant, teaching him how to spit shine shoes and mix a perfect highball cocktail. Under the big tent, Hamid became a circus sensation. At thirteen, he won a competition at New York's Madison Square Garden and earned the title of the 'world's greatest acrobat'. On the road from town to town, Hamid sat at the foot of another performer, the sharpshooting sensation, Annie Oakley. Years later, he would tell people that the big-top star taught him how to read and write. Yet behind the scenes, all was not well with Buffalo Bill's circus. The ringleader was drowning himself in booze, and eventually his illness cost him his business.

After Buffalo Bill folded his tent, Hamid drifted from circus to circus. In his twenties he formed his own show, 'Hamid's Oriental Circuses, Wild West, and Far East Show'. But Hamid could not make it on his own, not yet at least. Tired of the show-business grind, he tried his hand at the oil business. His Texas ventures came up dry and he drifted back to what he knew – the circus. Using his contacts, Hamid started to book tumblers, elephant trainers and clowns in circuses and other show-places across the United States and Canada. By the end of the 1920s, his growing business supplied more than 300 fairs and 35 amusement parks with their acts. Noting his success, North Carolina officials hired him to run their state fair. But they fired him a couple of years later after discovering that he was drawing a higher salary than the Tar Heel Governor. He quickly got hired to run the New Jersey state fair. That brought him within striking distance of Atlantic City.[6] In 1938, Hamid decided to take out a lease on the Million Dollar Pier, one of the city's three spots, influenced by the piers of Great Britain, which jutted out from the Boardwalk into the ocean and allowed visitors to go on rides and listen to live music.[7]

What Hamid really had his eye on, though, was Steel Pier, first opened in 1889 and the gem of Atlantic City piers, known in those days as the 'Nation's Showcase'. In 1945, owner Frank Gravatt put the amusement site up for sale. Legend has it that Gravatt thought of Hamid as a cheap carny man and refused to sell him his prized possession. Hamid fooled Gravatt, purchasing the Pier through the Abel Holding Company.[8] Whatever disagreements divided the two strong-willed men, Hamid continued what Gravatt had started, running the Pier as one-stop entertainment bazaar, part downtown, part carnival and part circus. He knew some people liked the movies more than the diving horse, the boxing cats ahead of the child stars, the big bands more than vaudeville. But he did not care. He wanted as many people of all ages to come through the front gates as he could cram onto the Pier, so he tried to offer something for just about everyone (Figure 7.6).

Like much of mid-twentieth-century America, Steel Pier practised segregation. More importantly, it created and layered, at the same time, white supremacy. Only women and men who could pass as white could get on the Pier at any time and do whatever they wanted before the 1960s. Yet, the presence and performance of race remained central to its mass appeal. Beginning in the 1920s, Steel Pier featured, for instance, an 'authentic' Hawaiian Village. Throughout the day, 'real' Hawaiians came out of their thatched-roofed huts at regular intervals to

*Figure 7.6* Hamid and his crew in front of a sign advertising vaudeville acts and minstrel shows, about 1945

Source: © ACFPL Atlantic City Heritage Collection

entertain the crowds. Dark skinned men in flowery shirts played bouncy island tunes, while dark skinned women in bikini tops and grass skirts shimmied and swayed in a version of the hula dance performed only on the mainland. Parents gawked and mumbled to their children, 'look, there's a real native'.[9] After watching the performance at the Hawaiian Village, some went to see that staple of mass American culture on the Pier, the minstrel show, just like they might have at the carnivals, fairs and local Kiwanis Club and fire department fundraisers in their hometowns.[10] Everyone, from Al Jolson to Bing Crosby to Sammy Davis Jr to Eddie Cantor, performed at some point during their long careers in black face. On the Boardwalk, Steel Pier presented minstrel shows well into the 1940s.[11] During these acts, 'coloured comedians' and 'black-faced singers' told mythic tales of the Old South, conjuring familiar images of shiftless plantation hands and hysterical nannies. Every performance included a couple of songs of yearning for the 'protective hand' of the 'dear ole' massa'. When the minstrel tales shifted to the north, they poked fun at the fatuous dandy and the uppity city slicker. Throughout it all, the white audiences at Steel Pier grinned and clapped. From there, they moved on the next event.[12]

'Nowhere else', a Pier brochure bragged, 'could you get so much, so good for so little money'. 'America's Great Amusement Bargain' was how Hamid's marketing team described his showplace.[13] Throughout the middle decades of the century, the Pier was in the business of numbers. Giving people a choice of three or four movies, 'crack bands', a water circus and a glimpse of the 'stars of tomorrow' for 'one low price' required, above all, a mass audience. As long as Atlantic City remained a mass resort, the Pier had its crowds and could entertain them at an affordable price. Fifteen thousand women, men and children flocked to the, 'Showplace of the Nation' on most summer days during the city's glory years. When it rained and people could not go to the beach or stroll the Boardwalk, the number jumped to 20,000. In 1939, a few years before Hamid purchased the Pier, almost 40,000 came to see Amos-n-Andy live and in person. Twenty years later, more 58,000 showed up to watch the fresh-faced, teen idol Ricky Nelson. That was the Pier's peak.[14]

Almost everyone who met Hamid commented on his restless energy. Up early in the morning and late to bed, the old tumbler worked like a Depression kid afraid that everything he had could be gone in an instant. By the mid-1960s, this feeling was not just an illusion. Things were slipping away from Hamid. The crowds on the Pier got smaller and, in his mind, more ragged each summer. While some Boardwalk business people moved to Florida, others decided to get what they could, while they could. They fleeced summer customers and late season conventioneers, pocketing the profits from overpriced lemonades and spaghetti dinners and putting little back into their Boardwalk stands and restaurants. Hamid, however, did not give up on the Pier or Atlantic City so quickly, though it was not like he had a business that could be easily packed up and moved to another state.

When a storm ravaged the city in 1944, Gravatt, the then owner of Steel Pier, quickly rebuilt the place and erected a gold-domed dance hall at the end.[15] When high winds and high tides blew through Atlantic City in 1962, pushing a runaway barge into the Pier and hacking a hole into its movie houses, Hamid reinvested in his showplace. Seven years later in 1969, a fire broke out, consuming a fifth of the Pier. Again, Hamid rebuilt, but this time he did not use any gold. The Pier did not generate the revenue it once did.[16]

'As tourists stopped coming to Atlantic City', Hamid commented, 'the numbers [for Steel Pier] declined. It got hard to run'.[17] Just like urban amusement parks in New York, Cleveland and Los Angeles, Atlantic City lost its mass crowd in the 1960s.[18] In response, Hamid worked on his showplace and worked on local politicians and city leaders to bring the masses back to the Boardwalk. Yet, like most leisure entrepreneurs, Hamid did not want just any crowd, he wanted what he judged to be the right crowd.

## The dilemma of the crowd

Hamid talked about the crowd all the time. In 1969, he spoke of the city's need to find 'decent people' to fill the streets of Atlantic City that the board game Monopoly had made famous and that had, in his words, 'deteriorated into ghettoes'. Other times, he talked about the necessity of luring, 'middle-class vacationers back to town'.[19] What Hamid proposed was the rebuilding and adding to the mass public of the past – the public that had gone on vacation with increasing numbers after the 1960s to the Florida coast and ex-urban and gated places like Six Flags and Disneyland. Hamid never spoke explicitly about the racial or class make-up of the 'decent' crowd. He did not have to – just about everyone knew what he meant when he talked about 'decent' people and 'middle class vacationers' – these were white people, men in blue blazers and creased pants and women in silk dress and cashmere sweaters. This was the crowd of Atlantic City's past. These were immigrants and storeowners, plumbers and accountants. They were people on their way up and out of the working-class. The sons and daughters of immigrants, they were people celebrating and showing off that they had made it America. What they all had in common was that they could pass for white. Atlantic City in its hey-day was just like Steel Pier. African Americans could perform there and play jazz and swing there and even appear in black face there. They could push families down the Boardwalk on rolling chairs and serve them in fancy restaurants, but they could not check into one of the city's showy hotels, or sit where they wanted at the movie houses, or sunbathe on whatever beach they wanted, or dance in the ballrooms with someone of a different race.[20]

It was these well-dressed white people, the respectable, patriotic strivers, who Hamid wanted to lure back to Atlantic City. African Americans, casually dressed day-trippers and ragged hippies could go somewhere else. The question was whether there was enough of the middle-class with the middle-class sensibilities that Hamid applauded left in America to sustain Atlantic City and Steel Pier. Perhaps the coming apart of America – something much on the minds of commentators and elected officials in the 1960s – had an impact beyond politics and music and fashion. It remade, and fragmented, commercial spaces like George Hamid's Steel Pier.

George Hamid Sr was hard to miss. By almost any metric or aesthetic standard, he was handsome. His face was shaped like a chiselled rectangle. His eyes looked like jewels, shining bright and clear, twinkling often with a touch of mischief. Tall and broad-shouldered, yet agile and graceful, he appeared well into his fifties like he could still anchor a circus pyramid or soar across the big-top on a trapeze. But it was his neatness that was his most striking feature. His pants were always perfectly creased and his shirts were filled with enough starch to withstand a windstorm and enough bleach to disappear into a white cloud. He always seemed to have come straight from the barbershop. His hair was perfect; Brylcreem provided whatever discipline repeated cutting and combing could not master. Hamid's moustache was as exact as his hair. Wide and tidy, it stopped just before the corners of his mouth as if to call further attention to its owner's fastidious good grooming. Outward appearance clearly mattered to Hamid. Good clothes and neat hair seemed to him, like they did to many middle-aged Americans in the post-war era, to have an almost moral quality.[21]

'Many former visitors', the members of the Atlantic City Women's Chamber of Commerce complained in 1959, 'stopped coming to the resort because the tone of the Boardwalk has been lowered'.[22] Hamid agreed. In the late 1960s, he railed against 'scantily clad strollers', 'people walking barefoot' and 'men without shirts'.[23] Rebuilding the public, Hamid believed, required re-establishing the formal and informal dress codes of the past. He pressed city leaders to enforce local statutes on Boardwalk attire, satisfied no doubt when city commissioners in 1967 passed an ordinance making it illegal for anyone to saunter down the promenade 'clad only in a bathing costume'.[24] The next year, two dozen 'prominent' local business and political leaders appointed Hamid's son, the handsome Princeton graduate who entered the family business in the 1960s, head of a committee charged with upgrading the Boardwalk's tone. Addressing the press, George Hamid Jr proclaimed, 'we are one of the proudest resorts in the world and we want to maintain our dignity and stature'. Along with enforcing dress codes, Hamid said his group would press for regulations on Boardwalk merchants and storefronts. His father echoed his son's concerns, urging city leaders to outlaw blaring loud speakers and garish window displays. A better-looking Boardwalk, Hamid believed, would bring 'better' people back to the city.[25]

The Hamids also knew that bringing back 'better' and 'decent' people meant doing something about crime, or at least the perception of crime. Certainly, father and son agreed with the President of the Atlantic City Improvement Association, who said in 1968, 'The Boardwalk should be the safest street in America'.[26] To give it that feel, and to reassure visitors, the elder Hamid pressed city leaders to install brighter lights and recruit more police officers. In 1968, the Chamber of Commerce, to which Hamid belonged, predicted that the changes would make 'people less fearful of muggings'.[27]

The owner of Steel Pier wanted to put one more confidence builder in place. Hamid spearheaded a petition drive calling on local officials to deploy K–9 police dogs on the Boardwalk. Some African American leaders opposed the idea. Civil

Rights advocate Manolette Nichols told city commissioners that he worried that the K–9s 'would be used against the black community for the protection of white people'. Alfred Washington, head of Atlantic City's Afro American Unity Movement, suggested that city leaders investigate the causes of crime instead of unleashing police dogs, adding that he thought there would be 'mass hysteria in the black community if police dogs were to be used'.[28] Hamid dismissed these objections – objections that revealed the level of racial distrust in the city – as little more than the cantankerous outbursts of professional activists and apologists. But Hamid probably did not care; he was not trying to win over the local black community. Lights and dogs, he trusted, would re-establish control over the public realm, giving the city a chance to win back the well-dressed, white middle-class crowd.

Hoping to turn the city around, the Hamids backed Atlantic County's tenacious and forceful political boss, Frank S. 'Hap' Farley. When the Republican leader, Farley, threw his political weight around and persuaded the Democratic Party to bring its national convention to the Boardwalk in 1964, the owner of Steel Pier praised the move. Like most everyone else in town, he thought that the convention would showcase the city.[29] Hamid expected a windfall as well. Just to make sure, he dressed the diving horses in LBJ coats and booked Eddie Fisher, Mickey Rooney and Milton Berle to perform in his splashy showrooms.[30] As it turned out, the horses and entertainers played to mostly empty houses. But it was the blistering press reports that troubled Hamid the most. 'This is the original Bay of Pigs', one newsman joked. Newspaper readers in Indianapolis and Baltimore learned about hotel switchboards in Atlantic City that did not work, room service that took hours to deliver cold food, and drink prices that soared just in time for the convention.[31]

Hamid knew, as Walt Disney did, that consumers' fantasies depended, in part, on service. Waiters, doormen and bus boys had to put visitors at the centre of the tourist narrative. Making them feel like kings and queens by indulging their dreams and desires generated return business. Taking a cue from Disney's theme park, where managers turned workers into cast members, Hamid and others from the Chamber of Commerce wanted to make sure, in the convention's wake, that service workers dressed neatly and treated guests as royalty. 'Bad impressions', it was argued, 'from a discourteous act take long to heal'. To help the resort win back visitors after the convention and other disappointments, Hamid recommended that waiters and clerks take 'a short course in courtesy'.[32]

Hamid copied from the Disney model – and other successful post-war ex-urban amusement parks – in additional ways. In 1969, he promised that if Steel Pier received 'the support of the merchants and hotel/motel people, we will add attractions like Disney does'.[33] When General Motors closed down its Boardwalk showroom in 1968 after more than fifty years of displaying cars by the ocean and paying six figures for the privilege, Hamid put a seaquarium in its place.[34] At the same time, he talked about adding a 'big zoological garden' with seventy-five animals that would combine 'entertainment and education'.[35] The changes marked a shift in Hamid's thinking. In the past, the showman presented animals as spectacles, as racing fish, boxing cats and diving horses. But he sensed – quite rightly – a change in middle-class attitudes by the late 1960s. 'Decent people', the kind he was trying to lure to the city, now imagined themselves as the kind of people who cared about seeing fish and animals in natural seeming habitats.[36]

Next to the aquarium and petting zoo, Hamid planned, around the same time, to build a full-sized ice-staking rink. All three of the Pier's new attractions, the circus man believed, would enhance the middle-class feel of his showplace and of the city. Nothing seemed to capture this better than the ice rink. Ice-skating still carried with it the kind of absurdity that made it an ideal Atlantic City attraction. What could more fantastic – and nature defying – than ice-skating at the beach in the middle of the summer? Only a truly leisured class of normally hard-working folk deserved such a delicious twist of fate. But clearly ice-skating also had a target audience. Hamid must have known that few poor people and few African Americans owned ice skates or would spend their leisure time twirling around freezing rinks in the middle of the summer. Hamid never talked in public about the demographic profile of the audience he was trying to bring to town but, like Fred Thompson and Walt Disney, he sought to shape his attractions to match the constantly shifting tastes and values of the white middle-class millions and he tried to segregate, as he once did publically, without calling attention to the racial filtering process.

Even though Hamid tried to make over Steel Pier to reflect the shifting self-perceptions of the late post-war white middle class, he never went far enough. His new attractions were imitations, but only pale imitations of the truly new and colossal attractions built, for instance, at San Diego's Sea World and the same city's award-winning zoo. In part, Hamid did not have the physical space or capital or corporate backing to construct and finance a massive nature-based theme park. Yet that was not the only problem. The high-flying circus acrobat stubbornly clung to the old crowd, those striving immigrants and their families on their way out of the working-class who crammed onto the Pier every weekend during the inter-war years. He knew that much of his business even in the 1960s ran on the fumes of nostalgia. Each year, tens of thousands of people still came to Atlantic City, and to Steel Pier, many of them to relive a bit of their past. Not wanting to lose a single one of these customers, Hamid continued to feature vaudeville-style acts, women tumbling on horses into pools of water, and swing bands long after the music disappeared from the Billboard charts.

Unlike Disney, Hamid and his son were tinkerers, not revolutionaries. They did not radically rethink the idea of Atlantic City. They still imagined the resort as a place where people came to act out their entrance into the nation's middle-class mainstream. That is why, well into the turbulent 1960s, Hamid Sr continued to fly the flag, turning his own Horatio Alger story into a central theme of the Pier's publicity. His journey from circus rags to Boardwalk riches to father of a Princeton graduate proved, he professed, that the American Dream worked. By coming to the Pier, he suggested, visitors paid homage to the nation and the ideal of social mobility, even in the face of Vietnam and Woodstock. But as the generational divide grew from a crack to a chasm, the stories the Hamids told and the attractions they mounted no longer seemed so benign and innocent of politics. Some read them as signals of the Pier's apparent allegiance to the older order. The legion of youth of the 1960s did not tolerate squareness and rejected any sign or whiff of it outright.

Yet, as much as Hamid wanted to bring back to the Pier the old days of well-dressed crowds of unified families, he was smart enough to know that the past could not be completely recreated. Surely he read the reports in the local paper about Atlantic City Club meetings *in* Miami Beach and of the Disney Corporation's plan to build a new theme park in Florida?[37] With many of the old faithful in the Sunshine State or Bermuda or at home wading in backyard pools or watching television in their air-conditioned living rooms, Hamid knew that he and other Atlantic City business leaders had to find new groups of 'decent' people to make up a re-configured mass on the Boardwalk.

Trying to capitalise on the post-war marriage boom, city officials, beginning in the 1950s, dubbed June 'Honeymoon Month'. Ten years later, feeling the first pinch of decline and looking to add to the crowd, the Hamids and others redoubled their efforts on this front. For a cut-rate price, honeymooners who came, in the mid-1960s, to celebrate their nuptials along the Boardwalk got a room, meals, sight-seeing trips, a box of salt-water taffy, a rolling chair ride and passes to Steel Pier.[38]

As much as the Hamids liked to watch newlyweds slow dancing on the Pier and holding hands on the Boardwalk, the father and son team were even more enthusiastic, at least from a business perspective, about Canadians. These women and men from the north represented just the sort of 'family tourist trade' – in other words, straight, white couples with kids – the Hamids wanted to see on the Boardwalk. 'The Canadians saved us from complete disaster last summer', Hamid declared in 1970, 'they still come with their families and stay . . . for a week or two and spend money all over town'.[39] To keep them coming and spending, the one-time circus promoter opened an office in Montreal. The city's tourist board and Chamber of Commerce followed suit, leasing storefronts in several eastern Canadian cities. Each year in late June, just as the honeymooners checked out of the hotels, local leaders hosted Canada Week. They started the festivities off with a 'Salute to Canada', followed by the crowning of 'Miss Canadian Visitor'.[40]

Canadians and honeymooners, as important as they were, the Hamids knew, could not save Atlantic City; they could stop the city's bleeding of tourists but they could not heal the wound. There were just not enough of them willing to be married again and again and make the drive every summer. To bring back the masses of decent people, the city, the Hamids and others seemed to think, had to put up new clearly marked walls and more effective gates to keep the poor, scruffy teenagers and large numbers of African Americans out of sight. Here, the law acted as both a tool and an impediment. City leaders could not in the 1960s revive the Jim Crow regime of the past or bring back police tactics of intimidation where officers shooed African Americans away from the Boardwalk and from 'white' beaches. Even if they wanted to bring back the past, the Hamids knew that African Americans and civil libertarians would not let them. Still, they knew enough about the white middle-class public to know that it could not take shape without exclusion. But again, the beach and the Boardwalk remained public places, legally at this point open to anyone. Unlike Disney officials, Hamid could not relocate beyond the city limits in order to keep the wrong people away – 'roving groups of hoodlums', in one man's words. The owners of Steel Pier had to find a way around legal barriers against discrimination to recreate what they envisioned as a viable and not entirely inclusive post-Civil Rights, urban middle-class public.[41]

The elder Hamid's strong support for dress codes and sanctions against tacky Boardwalk displays represented two methods of keeping the 'indecent' at bay. But the city, he thought, had to remain on guard on all fronts. The public, he knew, was precarious. Before 'the dawning of the age of Aquarius', Atlantic City operated as a family resort.[42] Culturally, the social world of the young closely mirrored the world of adults. Teenagers put on their best clothes before heading for a night out on the Boardwalk, they rode the rolling chairs if they had extra money, they went to the movies in pairs, and they tapped their feet to the snappy riffs of Woody Herman and the 'Rumba King', Xavier Cugat. Even the first stirrings of rock-and-roll did not disturb the peace between kids and parents on the Boardwalk. Bowing to market pressure, Steel Pier in the early 1960s booked acts like Ricky Nelson and Paul Anka. While some older women and men might have scratched their heads at the new electric rhythms, they knew that the smiling white boys singing about 'puppy love' were really good kids.[43]

In the late 1960s, Hamid could be heard grumbling about 'long-haired hippies'. When a music festival brought Janis Joplin, the Jefferson Airplane, the Byrds, Creedence Clearwater Revival and Frank Zappa to the Atlantic City Race Track, a dozen miles from the Boardwalk, a week before Woodstock in 1969, Hamid fumed 'these kids are . . . repulsive', adding that they would come 'without leaving twenty cents behind, unless, of course, they buy a hot dog and some pot'.[44] Given how he felt about the emerging youth culture, Hamid almost certainly supported measures proposed by local politicians aimed at keeping freak-flag-flying members of the counter-culture at a distance. He most likely backed plans to limit the number of people who could stay in individual hotel and motel rooms. No doubt he also agreed with local residents who called for the closing of the Psychedelic Fun House. A self-described 'living theatre', the gritty coffee house located on North Kentucky Avenue, on the African American side of the city's rigid racial housing divide, staged *Hair*-like musical revues several times a week. Not long after it opened, the Fun House applied for a liquor licence. One city commissioner opposed the move, describing the theatre as 'obscene' and charging that it appealed to 'a prurient interest' with its 'outrightly lewd and indecent' performances that lacked 'redeeming social value' and were 'patently offensive because [they are] affronts to contemporary community standards relating to sexual matters'. 'If I were to approve this license', he continued, 'I would in effect . . . be a participant in a criminal action'. Within a couple of weeks, the city commission received two petitions, signed by thirty-four persons, objecting to the Psychedelic House getting a liquor licence. It is not clear if Hamid joined these protests but he undoubtedly wanted to bar blue-jeans-wearing, long-haired hippies from the city and Boardwalk.[45]

Throughout the late 1960s, Hamid sometimes put day-trippers – 'shoobies' in the local parlance – in the same category as hippies. (Shoobies earned their nickname because in the old days they supposedly jammed all the stuff they needed for a day at the beach into a shoebox.)[46] Unlike long-haired kids, shoobies usually came to town as families, but to Hamid they were the wrong kinds of families. He complained about how they ate on the beach and changed in public restrooms or in the backseats of their cars. Mostly, however, he groaned that they did not spend enough money. Unable to afford a hotel or motel room, they drove in for the day, taking up precious parking spaces. They trudged off to the beach. Sometime in the middle of the afternoon, just as their skin started to turn bright red, they stopped off for a lemonade or ice cream cone, and then left town before nightfall. Although he never said it out loud, Hamid clearly worried that the shoobies in their dripping bathing suits, sand-matted hair and hot-dog tastes gave Atlantic City too much of a rough-hewed, working-class feel.[47]

To discourage day-trippers, Hamid pressed for a beach fee. Before this, Atlantic City had not charged visitors to use its beaches. Some thought this gave the city a competitive advantage over the beach towns to the north and south that required people to pay to sunbathe and body surf. Hamid wanted an admission price. He wanted to bring to the town the kind of people who would not think twice about paying a small fee to use Atlantic City's beaches. He thought this would act as a filter. Knowing that shoobies, much like him, kept a close eye on the bottom line, he thought they would take note of the beach fee and take their frayed beach towels and sack lunches to another shore town where access to the oceanfront did not cost anything.[48]

The Hamids worried even more, it seemed, about the Boardwalk's racial mix. By the late 1960s, segregation was, of course, falling apart in Atlantic City and most other urban places in the north.[49] As it did, the cramped, stuffy, un-air-conditioned rooming houses in the city's African American neighbourhoods went out of business as fewer black visitors accepted the crumbs of Jim Crow tourism. Taking advantage of new sets of choices, African Americans checked into hotels and motels closer to the beach. While the Hamids might not have welcomed the idea of African Americans staying at the fancy and ornate Marlborough Blenheim, they were, it seems, more troubled by black shoobies and bus tours filled with less affluent African Americans from nearby cities. To them, this was the wrong crowd.[50]

Following the legally mandated desegregation of Atlantic City's public places, African American church and civic groups sponsored regular bus trips to the 'Queen of Resorts'. 'On summer weekends', one journalist observed with telltale, and it turned out typical, signs of racial bias, 'busloads of blacks pour into Atlantic City from Philadelphia, New York, Wilmington, and other cities'.[51] Two Philadelphia reporters visited the Boardwalk in 1970 and commented, 'far more than half the people were black'. Most, they added, 'came down . . . for the day on a bus'.[52] 'Now', wrote another pair of journalists in 1970, 'the city has reached the point where the blacks have become too numerous for the comfort of white residents'. The overwhelming majority of Atlantic City black tourists, they said, 'look like everybody else, with perhaps flashier tastes in slacks and headgear'.

But it was black teenagers, roaming the Boardwalk wearing 'big sunglasses, enormous floppy caps and the 'in' thing this season, a version of a cowboy hat with the brim pulled down, hillbilly style', who really alarmed a number of whites. 'These kids', the journalists concluded, 'are only trying to be cool but they give white tourists a pretty good case of the creeps'.[53]

Hamid never commented directly on African American tourism.[54] Yet, a *Philadelphia Inquirer* reporter, who spoke with the Pier owner while working on a series on Atlantic City's decline, wrote, 'businessmen and public officials privately complain that this [the bus trips] has been hurting the white tourist trade, driving business to the smaller, quieter, whiter resort communities up and down the South Jersey coast'. In post-Civil Rights America, there could not be, it seems, both a black AND white, middle-class AND working-class tourist trade. It was largely one or the other. Hamid had to choose, and he chose the familiar world of yesterday over the unknown world of tomorrow.[55]

Again, Hamid never said anything in public about the buses and the bus riders but he and his son strongly supported an ordinance raising the charge on buses entering the city on Saturdays, Sundays and holidays. He pressed city commissioners, moreover, to pass a measure requiring all charter buses to leave town by 11 p.m. When local African American leaders objected to the recommendations, Hamid exploded, sounding like a man who had finally had enough. 'Let's face it', he fumed, 'the militant, loud, minority spokesmen' are not against the proposal 'for moral purposes'. They are against it, he wrote in a letter to the editor of the *Atlantic City Press*, because the new measures would stop their 'fellow travelers' who 'loot, steal, and abuse law-abiding citizens'. The city, he continued, 'should not be exposed to rowdyism whether it is done by long-haired hippies or Negro gangs or Mexicans, Asians, Indians, or whatever'. If these groups reigned over the city, Hamid warned, the 'beach, Boardwalk, and streets' would never be 'safe and decent', and 'there will be no tomorrow'.[56]

## Epilogue

Tomorrow did not really come to Atlantic City until after the casinos arrived. That is because Hamid was in the bind that most urban amusement park and city-based tourist entrepreneurs faced in the wake of the Civil Rights Movement. They could not keep people out anymore, and they could not create a place that was safe enough and dreamy enough to bring in a big enough crowd to sustain their massive, fixed and capital-intensive tourist complexes. In this way, Steel Pier was a lot like cities themselves in the 1970s. They were the past in an era of suburban succession. The tourist entrepreneurs who thrived at this time either worked out a way to attract a niche audience or built new places, with gates and walls, outside the city and away from the emerging diversity of the urban centres. Inside cities, the places that thrived resembled casinos, closed off and heavily guarded. They kept their crowds inside and at the tables, protected by cameras and uniformed officials. The crowds from the gaming halls did not drift off to Steel Pier between hands or pulls of the slot machines. Within five years of the opening of the first

Atlantic City casino, Hamid was gone and the Pier no longer mattered. Donald Trump wanted to use it to land helicopters for high rollers, not to entertain the masses. That was the logic for new tourist entrepreneurs in a new era. It would take another day, another moment, another couple of decades for Steel Pier to come back. But all the way back? That past was gone.[57]

## Notes

1　Harrison Rhodes, *In Vacation America* (New York: Harper and Brothers Publishers, 1915), p. 7.
2　For a model of studying tourist entrepreneurs see Woody Register, *The Kid of Coney Island: Fred Thompson and the Rise of American Amusements* (New York: Oxford University Press, 2001).
3　'Bridge to the Old World', *Time* (7 June 1961).
4　'Steel Pier Souvenir Program and Guide Book' (1954); Vertical Files, Atlantic City Piers, 2, Atlantic County Historical Society, Somers Point, New Jersey (hereafter cited as ACHS).
5　'Steel Pier Souvenir Program and Guide Book' (1949), Allen 'Boo' Peragment Library (private collection), Margate, New Jersey. See also Hamid's autobiography, George A. Hamid, *Circus* (New York: Sterling Publishing Co., 1950).
6　On the early part of Hamid's life see 'Bridge to the Old World', pp. 53–4; and 'George Hamid Dies at 75', *New York Times* (14 June 1971).
7　On the British influence see Steve Liebowitz, *Steel Pier, Atlantic City: Showplace of the Nation* (West Creek, New Jersey: Down the Shore Publishing, 2009), p. 39.
8　Liebowitz, *Steel Pier*, p. 62; *New York Times* (8 and 18 May 1945); *Atlantic City Press* (6 January 1980); Vertical Files, People, Businessmen, ACHS.
9　Pers. comm., Allen M. Peragment (5 and 12 September 2001); interview with Frank Havens by Cynthia Ringe (12 May 1978), Atlantic City Living History, Oral History Project, Heston Room, Atlantic City Free Public Library (hereafter cited as ACFPL). On the cultural interest in Hawaiians see Lewis A. Erenberg, *Steppin' Out: New York Nightlife and the Transformation of American Culture, 1890–1920* (Chicago: University of Chicago Press, 1981), pp. 224–6; and Jane C. Desmond, *Staging Tourism: Bodies on Display from Waikiki to Sea World* (Chicago: University of Chicago Press, 1999).
10　Ed Davis, *Atlantic City Diary: A Century of Memories, 1880–1985* (McKee City, New Jersey: Atlantic City Sunrise Publishing Company, 1986), pp. 99–100; 'Program for BPOE [Elks] Minstrel Show to Benefit the Welfare Fund' (15 April 1956), Ed Davis Papers, Folder, Davis Programs, Heston Room, ACFPL.
11　'Roving Reporter', *Atlantic City Press* (19 December 1943); Ed Hitzel, 'The Piers: From 1880 to Now', *Atlantic City Press* (18 October 1970). Hitzel notes that Steel Pier minstrel shows ended in 1945 as a result of 'the public's lack of interest'.
12　Liebowitz, *Steel Pier*, p. 49; Eric Lott, *Love and Theft: Black Minstrelsy and the American Working Class* (New York: Oxford University Press, 2013); Yuval Taylor and Jake Austen, *Darkest America: Black Minstrelsy from Slavery to Hip Hop* (New York: W. W. Norton, 2012).
13　'All Roads Lead to Atlantic City', *Atlantic City Press* (26 May 1963).
14　Liebowitz, *Steel Pier*, p. 59; Steven V. Cronin, '100 Years of Magic', *Atlantic City Press* (18 June 1998); 'Hamid Recalls Greats' (n.d.), Vertical Files, Steel Pier, Heston Room, ACFPL.
15　Interview with Peragment; also pers. comm. Lois Wallen (16 February 1999). On the storm see 'Heinz Pier: The Great Pickle Promotion', *Atlantic City Press* (16 May 1976); and Jim Waltzer and Tom Wilk, *Tales of South Jersey: Profiles and Personalities* (New Brunswick: Rutgers University Press, 2001), pp. 82–6.

16 On both 1960s storms see Davis, *Atlantic City Diary*, pp. 121–2, 130–31. Editorial, 'The Indestructible Pier', *Atlantic City Press* (31 December 1969); and Hitzel, 'The Piers'.
17 Cronin, '100 Years of Magic'; 'Hamid Recalls Greats'.
18 On the era's amusement pier crisis and the role of race see Victoria W. Wolcott, *Race, Riots, and Roller Coasters: The Struggle Over Segergated Recreation in America* (Philadelphia: University of Pennsylvania Press, 2012).
19 Michael Checchio, 'Campers Backed By Hamid', *Atlantic City Press* (2 March 1975); and 'Memo' (28 February 1975), File, Trav-L-Park, Box, Proposed Developers, R-115 – Marketing Campaign, Atlantic City Housing Authority, Atlantic City, New Jersey.
20 For more on this dynamic see Bryant Simon, *Boardwalk of Dreams: Atlantic City and the Fate of Urban America* (New York: Oxford University Press, 2004); and David Nasaw, *Going Out: The Rise and Fall of Public Amusements* (New York: Basic Books, 1993).
21 On the politics of dress and style in the 1960s see Kenneth Cmiel, 'The Politics of Civility', in David Farber (ed.), *The Sixties: From Memory to History* (Chapel Hill: University of North Carolina Press, 1994), pp. 263–90; and Gael Graham, 'Flaunting the Freak Flag: *Karr v. Schmidt* and the Great Hair Debate in American High Schools, 1965–1975', *Journal of American History* 91, 2 (September 2004), pp. 523–43.
22 'Women Push Legal Gambling', *Atlantic City Press* (17 March 1959).
23 'Merchants Complain about Encroachment Along the Boardwalk', *Philadelphia Inquirer* (24 August 1968).
24 'City Adopts 'Walk Tone Ordinance', *Atlantic City Press* (16 June 1967).
25 Frank J. Prendergast, 'Improved 'Walk Tone is Sought', *Atlantic City Press* (24 August 1968).
26 Greater Atlantic City Chamber of Commerce, 'Annual Report', *Action* (July 1958), Greater Atlantic City Chamber of Commerce, Atlantic City, New Jersey.
27 S. William White, 'TV Dogs Sought to Fight Crime on Boardwalk', *Philadelphia Bulletin* (22 December 1968). See also Minutes of the Board of Directors of the Greater Atlantic City Chamber of Commerce (25 January 1965), Greater Atlantic City Chamber of Commerce, Atlantic City, New Jersey.
28 'Hearing Re. Use of Police Dogs by A.C. Police Dep't' (2 January 1969), Minutes of Commissioners, Atlantic City, New Jersey, Atlantic City City Hall. See also 'Statement to the Board by Gregg Wells' (5 June 1969) and 'Statements Regarding the Establishment of a K–9 Corps' (18 December 1969), Minutes of Commissioners, Atlantic City, New Jersey, Atlantic City City Hall.
29 For an assessment of the convention's impact see Greater Atlantic City Chamber of Commerce, 'Demo Convention Meets Mixed Reactions', *Action* (September 1964), Greater Atlantic City Chamber of Commerce, Atlantic City, New Jersey.
30 Murray Raphel, 'The Democratic National Convention . . . Who Needs It? Atlantic City, Maybe?' *Atlantic City Press* (9 July 1972).
31 Theodore White, *The Making of the President, 1964* (New York: Atheneum Publishers, 1965), pp. 274– 5. See also Michael Pollack, *Hostage to Fortune: Atlantic City and Casino Gambling* (Princeton, New Jersey: Center for Analysis of Public Issues, 1987), pp. 21–3; Joseph F. Sullivan, 'AC: Images of '64 Made Blurry', *New York Times* (16 August 1981); interview with Pergament (3 March 1999). In a posting on 'Atlantic City Memory Lane' (iloveac.com.), Jim Bloom writes, 'The Atlantic City of the 1920s through the 1950s died after the Democratic Convention of 1964 and all the bad publicity. But the ghosts are still all around' (27 October 2000).
32 Tom Seppy, untitled, *Atlantic City Press* (2 February 1964), Vertical Files, History of Atlantic City Publicity and Promotional Materials, Heston Room, ACFPL. Greater Atlantic City Chamber of Commerce, 'Annual Report', *Action* (November 1957; March 1964), Greater Atlantic City Chamber of Commerce, Atlantic City, New Jersey; Recommendation for Greater Courtesy, R/UDAT – American Institute of Architects,

Regional/Urban Design Assistance Team, Atlantic City (November 1975), pp. 5–6, 10, Heston Room, ACFPL.

33  Untitled, *Atlantic City Press* (9 October 1969), Vertical Files, 'Steel Pier', Heston Room, ACFPL.

34  'The New Steel Pier Official Bicentennial Souvenir Program, 1976', Allen 'Boo' Pergament Library (private collection), Margate, New Jersey.

35  Untitled, *Atlantic City Press* (16 May 1968), Vertical Files, 'Steel Pier', Heston Room, ACFPL.

36  Susan Davis, *Spectacular Nature: Corporate Culture and the Sea World Experience* (Berkley: University of California Press, 1997), pp. 28–36; and Jennifer Price, *Flight Maps: Adventures with Nature in Modern America* (New York: Basic Books, 1999), p. 178.

37  Paul Learn, 'They Sing Praises of AC – But in Miami', *Atlantic City Press* (20 February 1966).

38  Interview with Mildred Fox by Cynthia Ringe (28 April 1978), Atlantic City Living History, Heston Room, ACFPL; Greater Atlantic City Chamber of Commerce, *Action* (June 1967), Greater Atlantic City Chamber of Commerce, Atlantic City, New Jersey; 'Special Rates for Honeymooners', 'Atlantic City Honeymoon Plan is Economy Deal Newlyweds', and 'Cut-Rate Honeymooners', *Philadelphia Bulletin* (16 April 1960; 23 May 1965; 15 May 1969).

39  James N. Riggio, 'Convention Priority Resented by Small Businesses', *Philadelphia Inquirer* (3 June 1970).

40  Ibid.; 'Walk Commission to Aid City Solons in Maintaining Tone', *Atlantic City Press* (7 June 1962); Henry Spier, 'Advertising Campaign in Canada Pays Off in Dollars for Atlantic City', *Philadelphia Bulletin* (15 August 1965); Greater Atlantic City Chamber of Commerce, *Action* (June 1966), Greater Atlantic City Chamber of Commerce, Atlantic City, New Jersey.

41  Frank J. Prendergast, 'Merchant Says 'Walk a Disgrace', *Atlantic City Press* (28 August 1971). See also Wolcott, *Race, Riots, and Roller Coasters.*

42  Line from the song, *Aquarius* from the musical *Hair*: http://www.metrolyrics.com/aquarius-lyrics-hair.html.

43  On the culture of teenagers see Michael Johns, *Moment of Grace: The American City in the 1950s* (Berkley: University of California Press, 2002), pp. 55–62.

44  Jim Waltzer and Tom Wilk, *Tales of South Jersey: Profiles and Personalities* (New Brunswick: Rutgers University Press, 2001), p. 38.

45  'Application of the Psychedelic Fun House' (15 July 1971; 26 August 1971; 23 September 1971), Minutes of Commissioners, Atlantic City, New Jersey, Atlantic City City Hall.

46  For an amusing, tongue-in-cheek article on shoobies see 'Delights of the Shoobie, or Atlantic City, on 15¢ a Day', *Philadelphia Inquirer* (9 June 1968); and Pollack, *Hostage to Fortune*, pp. 32–3.

47  'Merchants Complain about Encroachment Along the Boardwalk', *Philadelphia Inquirer* (24 August 1968).

48  Prendergast, 'Improved 'Walk Tone is Sought'. Years earlier, Hamid opposed such a measure, calling it a 'roadblock'; Dennis M. Higgins, 'O'Connell's Tax Plan Given Chilly Reception', *Atlantic City Press* (24 December 1962). Interestingly, some African Americans saw the campaign for a beach fee as discriminatory, see 'Beach Fee Considered in Atlantic City', *Philadelphia Bulletin* (21 September 1971).

49  For northern segregation see Thomas Sugure, *Sweet Land of Liberty: The Forgotten Struggle for Civil Rights in the North* (New York: Random House, 2009), and Matthew Countryman, *Up South: Civil Rights and Black Power in Philadelphia* (Philadelphia: University of Pennsylvania Press, 2007).

50 For a broader survey of race and beaches see Andrew Kahrl, *The Land Was Ours: African American Beaches from Jim Crow to the Sunbelt South* (Cambridge, Mass: Harvard University Press, 2012).

51 James N. Riggio, 'Boardwalk a Symbol of Black Frustration', *Philadelphia Inquirer* (2 June 1970).

52 Gaeton Fonzi and Bernard McCormick, 'Bust-Out Town', *Philadelphia Magazine* (August 1970), p. 58. See also Bruce Boyle, 'Of the Inlet Irish: It was Summertime and the Card Fell Right' and 'Technology, Racism, and Rolling Chairs May Revive Us Yet', *Philadelphia Bulletin* (22 December 1980; 23 September 1981).

53 Fonzi and McCormick, 'Bust-Out Town', pp. 122–3. In an interview with Murray Raphel and his wife, Ruthie, they also talked about white fears as a factor behind the city's decline.

54 'Crowd of Teen-Agers Dispersed at Shore', 'Two Boys Hurt in Teen Fracas at Atlantic City', *Philadelphia Bulletin* (9 and 10 September 1967); and Charles E. Funnell, *By The Beautiful Sea: The Rise and High Times of That Great American Resort, Atlantic City* (New York: Knopf, 1975), pp. 157–8.

55 Riggio, 'Boardwalk a Symbol of Black Frustration'. Business people in Coney Island echoed this assessment, blaming the decline of their resort town on desegregation: see Harold M. Scheck Jr, 'Coney Island Slump Grows Worse', *New York Times* (2 July 1964).

56 Letter to Editor, *Atlantic City Press* (5 June 1969).

57 Boyle, 'Of the Inlet Irish' and 'Technology, Racism, and Rolling Chairs May Revive Us Yet'.

# 8 The Parque de Atracciones de Vizcaya, Artxanda, Bilbao

## Provincial identity, paternalistic optimism and economic collapse, 1972–1990

*John K. Walton*

In the last years of the Franco dictatorship, at the end of what proved to be the final burst of prosperity for the metallurgical industries of Bilbao and district,[1] the Diputación or provincial government of Vizcaya, in the Basque Country of northern Spain, decided in 1972 to construct an ambitious amusement park at Artxanda,[2] a windswept mountain site above the provincial capital. Three regional banks provided financial backing, and work on the project began in 1973. The site opened, without fanfare and after two postponements, in mid-September 1974, at the very end of the summer season. It promised, and indeed delivered, the desired mixture of education, entertainment, relaxation and open-air leisure, employing 130 people at its peak, and apparently 'in its time was considered the most modern and best amusement park in Europe'. Its footfall, however, never lived up to expectations; the original projections anticipating 1.5 million paying customers a year, later scaled down to 1 million. But when the park closed in 1990, after several injections of additional capital from the Diputación during the preceding decade, and a full-scale re-launch as late as 1989 (at a time of serious economic depression), it was only attracting 120,000 annual visitors against a projected break-even figure of 640,000, even for the less ambitious of the two options for continued operation that were under consideration. Even at its peak, the park's annual footfall had never passed beyond (at most) half a million, so the prognosis was hopeless, especially in the straitened economic circumstances of the time, and in spite of considerable recent investment that had been aimed at turning the project round.[3]

Memories of the site remain a popular vehicle for nostalgia among those who were young during those years, in Bilbao and other parts of the Basque Country. The amusement park is commemorated online, in detail, on the website *Esperando al Tren* (Waiting for the Train), which records abandoned places that have the capacity to evoke buried memories and a sense of loss, and provides a basic timeline and description of the park's trajectory and attractions, together with an extensive pictorial archive and commentaries. The first part of the Artxanda compilation received 15,000 hits in its first three months, which is testimony to the lasting affection felt for a genuinely popular public attraction that was never able to live up to its grandiose economic ambitions.[4] This is only the most ambitious of several commemorative blogs and websites, and there have also been films and other artistic projects.

This chapter charts the trajectory of this unique venture, whose career spanned the troubled years of the transition to democracy in the Basque Country, investigates the reasons for its development and decline, and explores the cultures of commemoration that it inspired, setting the story in local, regional, Spanish and European context.

## Economic and political background

Bilbao was an interesting location for this project. It was the capital of the first Basque province to undergo intensive industrialisation and urban growth, both in Bilbao itself and along the estuary of the River Nervión, the Ría de Bilbao, accelerating dramatically from the 1870s onwards. Here, iron, steel and associated manufacturing industries had stimulated the explosive, unplanned growth of a string of smoky industrial towns along the west bank, whereby the 1970s the shipyards and factories had long ago swallowed up the last of the open spaces. The middle classes took refuge in the planned central *ensanche*, with its grand avenue and parks, or in a string of suburbs and seaside resorts that developed beyond the industrial belt, on the fringe of the growing city on the eastern side of the estuary. Overcrowding in working-class districts was often severe and sanitation lacking, but the nearby mountains provided outlets for those who enjoyed surplus time and energy, and by the early twentieth century urban entertainment was reaching out to the working class through music-halls, cheap theatres, dance halls, street entertainments and the cinema, as well as bars and brothels.[5] The population of Bilbao itself, the likeliest source of regular revenue for the Artxanda amusement park, had multiplied tenfold between 1877 and 1975, to reach 394,439 at the latter census. Since the 1960s, too, all but the most disadvantaged families had been enjoying some improvement in purchasing power after the privations of the Civil War and its aftermath, bringing more people within economic reach of the park and its pleasures.[6] Even so, if Artxanda had depended solely on the city of Bilbao, the original projection of 1.5 million annual visitors would have required almost four visits per year from every single inhabitant, regardless of demography. By 1975, the official total population of Bilbao and the towns along the estuary stood at 846,326, and that of the province of Vizcaya, whose government had promoted this venture, at 1,151,680. In retrospect, the scheme seems to have been based on highly optimistic assumptions, not least about future economic trends.[7]

The politics, and political history, of Bilbao are also relevant. It was the birthplace of Basque nationalism as an organised political movement. It was here that Sabino Arana incubated the ideology of what became, in 1895, the *Partido Nacionalista Vasco* or Basque Nationalist Party (PNV), inventing such symbolic accoutrements as the Basque flag or *ikurriña*, and the Basque national day *Aberri Eguna*, and giving his imagined nation the name Euskadi or Euzkadi. Arana's theories had xenophobic and even racist dimensions, and as the party gathered electoral strength through the early decades of the twentieth century it moderated its stance, shifting its emphasis to the politics of defending the Basque language and culture. This was far from being the only distinguishing feature of Bilbao politics – liberal, dynastic conservative and socialist traditions evolved from the later nineteenth century alongside Arana's nationalists – but its cultural influence persisted under the Franco regime, and was still there to be reckoned with as the brutal repression of the immediate post-Civil War decades began to fade.[8]

The Spanish nationalist project of the military rebellion against the Second Republic in July 1936, which brought Francisco Franco to power until his death in 1975, had no time for Basque pretensions towards nationhood. For this reason, most Basque nationalists fought against Franco, despite the strongly Catholic and Christian Democrat leanings of their party, and were thoroughly repressed in defeat. As a new post-war generation came to the fore, the authorities of the Franco regime strove to keep the lid on a bubbling cauldron of nationalist sentiment and mounting agitation, now expressed less through the Catholic Church than through trade unions, popular culture, the *ikastola* movement of schools, which taught through the medium of *euskera*, and increasingly the co-operatives of Mondragón.[9] All this was coming to the boil, accompanied by the long-delayed arrival of rising living standards and expectations, at the point when the amusement park proposal was being advanced. It was accompanied in the late 1960s and early 1970s by heightened labour unrest, inflation, political tension and repression, which only with hindsight could be identified as the dying throes of the dictatorial regime.[10]

There was, indeed, probably a strong element of 'bread and circuses' about this initiative. It is interesting to compare the Artxanda project with San Sebastián's official approach, in nearby Guipúzcoa province, to the visibly growing demand for popular tourism from its hinterland at about the same time, which was threatening to congest the beaches of this still-fashionable summer resort and damage its dominant model of up-market coastal tourism. Here, proposals were put forward to invest in new swimming pools on the urban periphery, with a view to filtering out popular demand before it reached the beaches; to provide new popular attractions in the interior of the province; and to convert the fishing harbour into a marina. These were not carried through, however, not least (in some cases) because of an unwillingness to provoke discontent. Conspicuous by its absence was any specific mention of a new or improved amusement park.[11] Similar ideas about channelling and managing demand, and an innovative and expansive role for government intervention in the leisure sphere, were thus at work in the distinctive environment of San Sebastián, where authoritarian instincts were tempered by recognition of the need to obtain the day-to-day consent of the governed; but in the Basque setting Artxanda was a purely Vizcayan conceit.

The Artxanda scheme was certainly aimed at promoting a sense of well-being, and perhaps even gratitude, which might bolster the shaky legitimacy of a crumbling, though still powerful, regime. An important context for the project was the decade of heavy investment in sporting facilities to which the provincial government of Vizcaya proudly drew attention, through two celebratory publications, in the year of the park's opening.[12] The park also promoted healthy and wholesome, if somewhat frivolous, leisure activities for families in the open air, in striking contrast with the disreputable urban pleasures of Bilbao and environs, which were enumerated in salacious detail at this very time in the *Guía Secreta de Vizcaya*, one of a series published in the mid-1970s, which took advantage of the moral relaxation of the end of the Franco regime to detail all the opportunities for indulgence in alcohol, illicit sexual activities and disreputable entertainment across Spain's cities and provinces.[13] Viewed from this perspective, the Diputación's amusement park looks like an exercise in promoting (relatively) rational, and certainly harmless, recreation through the provision of counter-attractions; an agenda that would not have been out of place in industrial England a century previously.[14]

It may also be no coincidence that the Artxanda project coincided with the growing prominence of ETA, the militant wing of Basque nationalism, which resorted increasingly to political violence or 'armed struggle' during the later years of the Franco regime, with a campaign of bombings, kidnappings and assassinations whose legacy is only now being tackled. This was gathering momentum during the 1960s and early 1970s, with considerable clandestine activity among the Artxanda mountains themselves, but it was still far from the grim levels of intimidation and carnage that built up subsequently, although the assassination of Franco's intended successor, Admiral Carrero Blanco, in Madrid in 1973 carried a particularly high profile. As a naïve attempt to outflank or discourage militant nationalism, however, the Artxanda project would have to be regarded as a failure; but no evidence has surfaced that any such imagined outcome, in a direct cause and effect sense, was in the minds of its promoters. The balance between paternalism, distraction and control in the minds of the Diputación and its collaborators would be hard to gauge even if direct evidence were brought to bear; but the general political circumstances do need to be borne in mind. The adverse economic conditions which came to the fore later in the decade could not readily be foreseen when the plans were made.[15]

## Artxanda and other amusement parks

The Artxanda venture was far from being the first modern amusement park in Spain. The one at Tibidabo, on a hilltop overlooking Barcelona, opened as early as the end of October 1901, not long after the pioneering ventures at Coney Island and Blackpool. Access from the city was provided by electric tramway and funicular, contemporary innovations in transport technology which often went hand in hand with amusement park development.[16] There were also early amusement parks in the Basque Country itself, beginning in the early twentieth century in and around San Sebastián, which was the leading Spanish seaside resort until the Civil War and beyond, and the capital of Guipúzcoa province, which adjoined Vizcaya to the east. San Sebastián acquired three early amusement parks. The first opened in 1902 on the coastal hilltop of Monte Ulía, reached by a tramway which linked up with the town's established urban network, and from 1907 offering the world's first cable car ride, the brainchild of the inventor Leonardo Torres Quevedo. A more conventional park at Martutene, originally based on some show caves, assembled its first rides from 1907 onwards; and five years later a funicular railway to the coastal summit of Monte Igueldo opened out the most enduringly successful of these ventures, which out-competed its rivals and is still in operation.[17] There were therefore plenty of regional precedents for the Artxanda project, in a highly visible tourist location. A more chronologically immediate but geographically distant Spanish precedent, which not only served as partial inspiration for the Artxanda park but was also a component of its founding company, was the *Parque de Atracciones de Madrid*, which opened to the public on schedule on 15 May 1969, offering thirty rides from the very beginning. The opening ceremony was attended, officially, by the entire municipal corporation of the capital, and entrance fees were priced very cheaply to encourage popular patronage. This proved to be a highly successful and stable venture, sustaining a regime of perpetual innovation and investment.[18]

Nor was the Bilbao amusement park unique in its short life, eventual abandonment, and capacity to generate affectionate nostalgia. A parallel example might be the Kulturpark Plänterwald in former East Berlin, also founded under an authoritarian regime, the German Democratic Republic, in 1969, although it survived the demolition of the Berlin Wall and, as Spreepark, continued in existence until 2001.[19] But what makes the Artxanda venture particularly interesting is a combination of its location, the timing of its opening in local and regional context, and above all the auspices under which it was promoted. Local authorities often lived in tension with their amusement parks, and when governments with a wider geographical remit deigned to notice them, their motives were perhaps more likely to be hostile than benevolent. Open-air industrial museums promoting a vision of regional popular culture, like the one at Beamish in north-east England, might be a different matter, though themselves capable of generating controversy. However, the tensions between Blackpool's municipality and its Pleasure Beach amusement park during the 1920s and 1930s, or the efforts of New York's planning commissioner Robert Moses to undermine the pleasure economy of Coney Island (while promoting 'respectable' beaches elsewhere), serve to indicate that the present case study of the promotion of an amusement park by a provincial government, which then made unrelenting if unavailing efforts to stimulate it and keep it in operation, is an unusual example worthy of analysis.[20]

## Creating an amusement park

In economic terms, the timing of this project could hardly have been worse, although this was not apparent when it was first proposed.[21] The old industrial base of Bilbao, founded on iron, steel and shipyards, went into catastrophic decline from the mid-1970s onwards, with rapidly increasing levels of unemployment. Between 1973 and 1983 the Basque Country lost nearly 28 per cent of its industrial employment, in striking contrast with the growth of 39 per cent that had been enjoyed between 1960 and 1973. Official unemployment levels stood at 1 per cent in 1973 and 20 per cent ten years later. The heavy industries of Bilbao and district were the worst sufferers, and the population fell while immigration went into reverse. The institutions of government were slow to react at all levels, which may help to explain persistence with the Artxanda project into and through the disastrous 1980s.[22] Meanwhile, the increasing freedom of expression associated with the post-Franco transition to democracy, in contrast with the repressive policies of the last years of the regime, brought the lack of urban planning, together with the defective housing standards and widespread dereliction of this industrial centre, port and provincial capital into even starker relief; and matters were not helped by the complexity of the Basque political structures and rivalries between the competing parties. Discussions on how to re-plan the city and its hinterland, and how to promote a transition from the secondary to the tertiary sector, were already under way in the second half of the 1970s; but serious debates over regeneration strategies were only just beginning in 1989, when the financial failure of the Artxanda project was at last becoming apparent.[23]

Under these circumstances, the use of public money to support and subsidise an amenity of this kind may seem surprising. As a contemporary analyst pointed out, the finances of the Basque provincial governments were problematic, while the general development of public services was inadequate and lagged behind other parts of Spain. Indeed, Basque tax revenues were siphoned off to Madrid and to other Spanish regions. Spending on social security, health, education, environment (especially river pollution) and transport (despite congested roads) was far below the national average, and a survey in 1975 showed relative dissatisfaction with all public services except electricity supply, refuse disposal and markets. The provincial governments of Guipúzcoa and Vizcaya provinces had been particularly sluggish and deficient in responding to rapid growth in population, personal incomes and employment over the previous fifteen years. The 1975 survey found that the aspect of public services that attracted most dissatisfaction was nature conservation, followed closely by parks and public gardens; and here at least the Artxanda project might seem to respond to public concerns. But there were many more pressing needs across the industrially blighted Vizcaya province, with its insanitary and overcrowded urban slums, than a new amusement park.[24]

The origins of the amusement park proposal reflect an authoritarian paternalism that also recognised a need to respond to the rising living standards of the later Franco period, after the 'apertura' or opening of the regime to external contacts, economic development and an incipient relaxation of religious constraints from around 1960 onwards.[25] The Vizcaya government's launching statement, in the formal and portentous language of the time and the regime, expressed official recognition of the province's need for:

> a place which, open to the four winds and to all kinds of people, might serve as a place of solace and freedom, recreation and relief in a life of hard work and commitment, for adults and children, providing installations which articulate the classical with the modern and might offer the cleanest and most natural medium for enjoying leisure; an environment which, at once unknown and desired, might attract the attention of everyone.[26]

This was, then, in its ostensibly extra-political way a democratic, inclusive ideal, which the regime was capable of articulating provided that what was on offer was controlled and did not present political, religious or cultural challenges to authority. Aspects of the regime's relationship with football run parallel here.[27] We shall see that the democratic rhetoric associated with the park was qualified in practice by the cost and inconvenience of access; and nostalgia for the amusement park tends to be expressed by those who were children when it flourished, although some of the rides were exciting experiences for adults, and the concerts and swimming pool also reached out beyond the patronage of children and their families.[28]

The park's creation entailed impressive civil engineering works, and up to 270 workers from the *Edificios y Obras* company were employed at a time. 300,000 cubic metres of earth had to be moved, and other indicative statistics include the laying of 22 kilometres of underground electric cable and 11 kilometres of underground pipes for water supplies and sewers. The Switchback (*Montaña Rusa*) had over half a kilometre of track, and all this added up to a major investment in public works for which the *Parque de Atracciones de Vizcaya S.A.* (Vizcaya Amusement Park Ltd) was constituted as an umbrella organisation, putting together capital assembled by the banks and by the *Parque de Atracciones de Madrid*, which was involved in the scheme from the very beginning. In practice, then, this was a public–private partnership. The original cost of the project was put at 515 million pesetas, which in October 2010 was equivalent to 3.1 million euros.[29]

## Attractions and identities

The park's architecture adopted the Brutalist style that was fashionable at the time, and was perhaps particularly suited to authoritarian regimes, although this did not prevent its appearance in other leisure settings.[30] The repetitive geometry and crude exposure of concrete mouldings in its built environment did not prevent the park's customers from identifying it with pleasure and excitement. It was designed by two Bilbao architectural practices, one of which, the Ortega brothers, had been responsible for the Bilbao headquarters of Bankunión, one of the three banks that supported the project. Its signature buildings were the eight reddish-orange pyramids, with steel superstructures on concrete pillars, which sheltered most of the attractions, and the imposing (indeed rather menacing) tower that housed the offices, bearing a strong resemblance to an airport control tower.[31] In this respect and others, the architecture of the park bears a startling resemblance to the nuclear power station at nearby Lemóniz, on the Vizcaya coast, an almost exactly contemporary project that became a symbol of externally imposed threat and danger, and was aborted when close to activation after radical nationalist campaigns of sabotage and assassination.[32] The contrast between the public receptions accorded to the Artxanda and Lemóniz projects, and the parallels between the projects which are graphically illustrated by their photographic archives, which capture their sudden abandonment and enduring decay, are arresting; but they are never discussed in the same moral universe.

The park was Brutalist, monumental and domineering, but also Basque. Until 1981 its official emblem was Chimbo, a small, wide-eyed bird carrying a flower in its beak.[33] The name *chimbo* (in the Basque language, *tximbo*) was applied generically to a dozen related species of warbler. It was identified historically with the citizens of Bilbao, who had hunted it as a delicacy in the autumn when it fattened up on ripening figs before migrating southwards, and for whom it became a nickname. This choice of symbol was, perhaps, in keeping with the subordinate status of the Basques under Franco, especially as the Spanish form of the name was used; and it was also identified with the locality rather than with the Basque Country more broadly defined.[34] For the Basque author Jon Juaristi, a few years later, the *chimbo* became a political bird, a symbol of the invention of middle-class tradition in the Bilbao of the late nineteenth century through the celebration of a rural, nostalgic political identity, associated with the promotion of a distinctive local dialect of Spanish rather than the very demanding mastery of *euskera*, the Basque language that was enjoined by Arana's nationalists. It was, then, logically representative of an alternative political identity, civic rather than nationalist, which was potentially available for incorporation into a Francoist vision of the Basque Country. Juaristi's polemic did not appear until 1994; but any earlier identification between the *chimbo* and his suggested alternative version of *Bilbaíno* identity would make the bird a suitable symbol for the Artxanda project which, by Francoist definition, identified Vizcaya province as integral to the traditional Spanish nation, rather than as part of a separatist Basque Country.[35]

The little bird's replacement was much more assertively Basque, in the powerful form of Basajaun, Lord of the Woodlands. This was a mythical benevolent bearded hairy giant, dwelling in the impenetrable depths of the forest. He dispensed practical wisdom to humanity, and was the friend of the shepherds, looking after their sheep and warning of wolves or impending storms. He lived in the forests on the flanks of Gorbea in Vizcaya, the highest mountain between Bilbao and Vitoria, the capital of Alava province; but the myth-making imagination could also find him around Ataun, in the wildest part of neighbouring Guipúzcoa, and in the Pyrenees around Irati, in Navarra. He was therefore a genuinely Basque figure, recognisable across and beyond Vizcaya province; and his avatar in the park went around distributing gifts to children.[36] The transition from Chimbo to Basajaun marched in step with the transition to democracy (and a new kind of modernity), and with the introduction of Basque rural sports to the park, effectively coinciding with the establishment of the Basque autonomous region, which was achieved in 1982.[37]

The attractions offered by the park were numerous, and innovations continued to be introduced until the end of its life, although the original rides survived untouched (and in increasingly faded state) until the very end. The entrance was through a flower-lined *Gran Avenida* (Grand Avenue), 250 metres in length, from which the various rides, amenities and services radiated. Luminous fountains reflected the current state of the art in such provision. The Switchback was a particularly exhilarating experience, for adults as well as children, as was the 26-metre *Noria* or Big Wheel, both of which provided spectacular views over the city and beyond. The *Pulpo* or Octopus was an exciting novelty in its time. There was also a gently paced miniature railway, and a tempting metaphor for the fall of Francoism was the replacement of a military 'fort' (with wishing well) by the swings and slides of a children's playground. Later additions included the

Scanner in 1980 (a spin-off from the film *Battlestar Galactica*) and an inclined wheel that was introduced in 1986 and derived its name from the spaceship *Enterprise* of *Star Trek*. There were also larger, permanent versions of the kinds of rides and stalls that would be familiar from the travelling fairs that visited Bilbao and smaller towns on their annual circuits: dodgems, roundabouts, a ghost train (literally a witch's train or *tren de la bruja*), trampolines and the swings and slides of ordinary children's playgrounds. Stalls offered games such as three darts for (at the end) 100 pesetas, to win small prizes. There was the *Casa Magnética*, a fun house with distorting mirrors; a house of fantasy with a Snow White theme illustrated by dolls, with a fearsome witch and an aesthetic far removed from that of Disney; and a *Casa Encantada* or enchanted house, full of fantastic imaginary creatures. There was a mini-zoo, complete with wolves and lion cubs, together with an aquarium and aviary. A 4,800-seater open-air auditorium defied the elements, and pop concerts were held there: the intention to provide a roof never came to fruition. A go-kart track proved attractive to adults as well as children, and there were other participatory car-based amusements. Four years after the general opening, a swimming pool was added to the attractions, although it had its own separate pricing regime. At the very end of the park's life, as part of what turned out to be the final push for reinvigoration, an area of slot machines gave way in 1989 to the *Selva Mágica* or Magic Jungle, an innovative attraction that offered computer-controlled games with animals. It was hired in under an expensive agreement, only to be destroyed by fire within a month of installation.[38] This was, then, an extensive complex that offered a wide range of attractions, from the basic and familiar to the complex and highly technological, as well as providing sites for entertainment, exercise and family picnics. It was a major operation.

## Problems and shortcomings

Despite heavy investment and a sustained commitment to innovation, the Artxanda venture laboured under several disadvantages from the beginning, and repeated complaints were made about aspects of its operation. These difficulties provide part of the background to the eventual collapse of the park's finances, and to its sudden closure early in 1990, in a precipitate abandonment of the site, which left behind a multitude of tangible traces to be treasured a generation later by the votaries of nostalgic commemoration.

Perhaps the most damaging and lasting problem suffered by the Artxanda site was its difficulty of access, whether by private car or public transport. It occupied ten hectares on the exposed slopes of Monte Ganguren.[39] This formed part of an extensive range of hills, with spectacular views over Bilbao and to the north, which since the late nineteenth century had been a place of escape and entertainment, fresh air and exercise. The hillsides were dotted with grill restaurants and *txakolis* or bars serving local wine. From the autumn of 1915, one flank of the range had been made more accessible by the opening of a funicular from the city centre, which was based in part on the similar enterprise at Monte Igueldo in San Sebastián, and was also intended at the time to service a projected amusement park on the Igueldo model. This did not happen, although the funicular terminal and its adjacent public park attracted a nucleus of cider houses, bars and restaurants; but the funicular survived, drawing some of its traffic from the pastoral farming on the hillside, which sometimes included calves being taken down the hill to the municipal slaughterhouses. The popularity of this area was boosted by the opening in 1975 of a new sports complex, the Ciudad Deportivo de Artxanda, which benefited from heavy investment by the Bilbao municipal government, which was also trying to provide healthy outlets for popular enjoyment in a way that might be thought to compete with the provincial government's amusement park. In 1975, the *Guía Secreta de Vizcaya* mentioned these developments, while commenting on the magnificence of the view and the importance of Artxanda as a place for romantic assignations, and listing five restaurants and wine bars near the upper terminus of the funicular; but, curiously, it made no mention of the newly opened amusement park.[40] Between 1976 and 1983, in any case, the funicular was out of action after a serious accident, and it had to close again for several months in the aftermath of the terrible Bilbao floods of the latter year. But it was never a relevant means of access to the Artxanda amusement park. This was a considerable distance from the funicular terminal, far beyond the range of pleasure-seeking families on foot, and suffered severely from the lack of a similar facility. Indeed, when operational the funicular should be regarded as a competitive alternative to the amusement park, providing a cheap and attractive local destination for Bilbao families on sunny afternoons.[41] This was, anyhow, a district that combined a range of characteristics, and there was fierce complaint during 1974–75 about a rubbish tip on one of the slopes of Artxanda, which was said to generate huge toxic clouds and make life very uncomfortable for Bilbao citizens when the wind was in the wrong direction.[42]

The road that actually led to the amusement park remained inadequate, as did the bus service and car parking facilities. The problems experienced on the day of the park's inauguration had lasting repercussions. The access road was narrow, winding and poorly surfaced, and the traffic jams on the belated opening day, with delays of an hour on both arrival and departure, were not calculated to encourage repeat patronage. The bus service was promoted as the preferred option but its timetable was sparse and its punctuality unreliable, as the buses themselves were caught in traffic, while the fares were almost as expensive as the admission charges for the park itself. The seven-storey car park had more than a thousand parking spaces, but it seems likely that planners' assumptions about the growing importance of car ownership were neither fully borne out in the structure of effective demand, nor followed through in the provision of infrastructure.[43] Similar issues surfaced in the early years of the industrial museum at Beamish in north-east England, which had similar democratic pretensions but was even less accessible by public transport.[44] Meanwhile, the new motorway from Bilbao to the east was already, in 1974, luring what motorists there were to the rival attractions of the beaches of Zarauz and even San Sebastián.[45]

The site was also terribly exposed to the elements, and the unreliable Biscay weather was sometimes too much for the shelter provided in this exposed location.

On the most inclement days, strong winds and driving rain ensured a soaking even for those who came well prepared. Mist and low cloud also presented regular problems. Unlike the consortium that investigated the possibility of setting up a home-grown British version of Disneyland in Blackpool in the early 1960s, only to abandon it as unviable after seeing the weather records, the planners of Artxanda seem not to have taken the Atlantic climate into account. It was, after all, very different from that of Madrid.[46] Between 1975 and 1979, the park was actually open throughout the year, on a daily basis between March and September, and on Saturdays, Sundays and public holidays throughout the winter. This optimism proved unsustainable, and even in the summer months the site was sometimes impossibly inhospitable. The normal 11 a.m.–7 p.m. opening hours were always vulnerable to inclement weather.[47]

There were other sources of complaint. Adverse comment appeared in the press about high food prices at the park outlets, which generally offered menus at the bottom end of the market: *churros* (fried dough strips), candy floss, ice cream, *bocadillos* (filled baguettes) and the inevitable pork steaks or roast chicken with chips. There was a *restaurant de lujo* (luxury restaurant) for parties and special occasions, with spectacular views and a capacity of 450, but it is not clear how well-used this was.[48] Even the ordinary admission prices might be a problem for large families or those on small incomes, especially (as noted above) when bus fares were factored in. The original 1974 admission charge of twenty pesetas for adults and ten for children might soon spiral out of control, especially as going on all the rides would increase the price for each child tenfold. In 1989, after fifteen years of inflation, a 500-peseta all-in ticket still did not cover the *Selva Mágica* or the go-karts, which carried a 100-peseta surcharge (50 pesetas for children). When food, drink and sundries were added, this was not a cheap day out, and the combination of cost and inconvenience was calculated to discourage multiple repeat visits. A trip to Artxanda for most families would be an occasional treat, indeed an expedition, perhaps annual at best, rather than the regular event the planners must have envisaged to arrive at their visitor number projections.[49]

## Decline and fall

When it came, the collapse of the Artxanda amusement park was sudden and dramatic; and it followed hard on the heels of a series of financial reconstructions and rescue bids which, rather than ringing alarm bells, may have sent out misleadingly reassuring signals about the provincial government's determination to keep the park open. In 1981, an additional capital injection of 300 million pesetas (nearly two million euros) was provided, adding roughly 40 per cent to the original investment; but this attempt at providing further impetus coincided with the worst phase of the industrial and financial crisis which accompanied the transition to democracy in Vizcaya. The new investment helped to keep the park afloat until 1988, when the provincial government brought 77 per cent of the shares in-house at an expenditure of 144 million pesetas or 900,000 euros, suggesting a sharp fall in the valuation of the enterprise over the past seven years. But the declared intention behind this regional municipalisation was not just to maintain, but to reinforce and expand. A year later 50 million pesetas were made available for a re-launch of the park, with an expensive advertising campaign and free buses from the centre of Bilbao. It was at this point that the single ticket covering (almost) all the attractions was introduced, and the expensive new *Selva Mágica* was brought in. The labour force, perhaps understandably, showed no inkling of impending crisis, and the summer of 1989 was disrupted by a series of strikes and demonstrations (sit-ins, hunger strikes, lightning strikes during the working day, leafleting of visitors) in pursuit of higher wages and parity of working conditions and benefits with established functionaries of the Diputación. But after this final, frenetic burst of activity and conflict, during which management seemed to lose heart and, in a contradictory stance, the Diputación kept delaying the 1989 season's opening to save money, the end came quickly. Losses of 188 million pesetas (1,130,000 euros) in the summer had to be borne by the provincial government, tucked away in a special section of the budget, and at the same time the damning results of the viability study emerged into daylight. The Diputación, and the two remaining banks (*Bankunión* had disappeared as a separate entity during the crisis of the early 1980s),[50] pulled the plug on Artxanda, and the whole complex closed, out of season and effectively overnight, at the beginning of 1990.[51]

## Afterlife

The apocalyptic nature of the end of Artxanda, whose sudden closure left an almost Marie Celeste scene of abandoned artefacts, scattered tickets, posters, menu boards, documents and files, and deserted rides and restaurants, was conducive to the development of a culture of commemoration and legend. This really gathered momentum in the new millennium, as the children who had enjoyed Artxanda, even if their experience was confined to a single, isolated, memorable visit, grew to adulthood and maturity, had children of their own and began to reflect on their past and their family histories.

The amusement park itself was left to decay, and acquired the romantic aura that goes with rust and invading nature, and inspires the souls of industrial archaeologists. Many of the rides were sold on to a Portuguese amusement park. Parts of the site were put to uses that could be set against the 200,000 euros per year that were spent on security and basic maintenance. The Diputación used it to store its superannuated buses, tractors, traffic lights and road tankers, and (ironically) rented parking and storage space there to itinerant fairground operators on their visits to Bilbao. But access for the general public was denied by security guards with their dogs.[52]

Under these circumstances the site developed its own mystique, as a decaying monument to lost pleasures. Seventeen years after closure, the artistic project *Vuelven Las Atracciones* (The Amusements Return) brought parties of nostalgic visitors to Artxanda on the autumnal weekends of October 2007. The artist Saioa Olmo set the project up, and her team presented it for several nights in mid-October, in the square opposite the Teatro Arriaga in the old centre of Bilbao. The special guided tours took 500 people to tour the decaying ruins, and left a long, unsatisfied waiting list. In March 2008, the artist Guillermo Santamaría brought out a commemorative video, and in 2010 a book of photographs and memories was published. There was no shortage of media attention; and sponsorship came, perhaps with an element of irony, from the Diputación Foral de Bizkaia as well as the Basque autonomous government and the provincial government of neighbouring Gipuzkoa.[53] This was the tip of a wider iceberg of informal commemorations on blogs and websites (most obviously the extensive coverage orchestrated by Tomás Ruiz on *Esperando al Tren*). The park also has an unusually full, detailed and well-documented Wikipedia entry.[54] The remembrance of Artxanda drew out not only the memories of those who had enjoyed the park, but also the wistful imaginings of those who had never been there, or were too young to remember their childhood visits. Here, as elsewhere, a site of lost pleasure, childhood and fun became an emblem of celebratory individual and collective memory, whether or not the celebrants had ever enjoyed the park's delights themselves.

## Conclusion

This was a unique venture. There was nothing remarkable about the content of the Artxanda park: even its Brutalist architecture had counterparts elsewhere. It was neither the only such project in the Spain of its time, nor the first of its kind. Its location on a hill-top overlooking an industrial city was unusual, and undoubtedly contributed to its failure to achieve the (optimistic) projected footfall, and to its short life and sudden abandonment. But what really stands out, in comparative context, is the role of the provincial government in imagining, adopting, enabling and eventually funding and sustaining a venture of this nature, beginning in the declining years of a military dictatorship, and continuing through a difficult period of economic, political and cultural transition until the economic situation suddenly became untenable at the beginning of 1990. An intervention of this type, on this scale, and from this level of government would have been unthinkable, for example, in Britain. Here, municipal authorities had a long track record of establishing, embellishing and maintaining public parks of a more conventional kind, sometimes including quite elaborate amenities and technologies for children's play or adult exercise (rowing, swimming, football), while central government had overseen the development of a kind of metropolitan theme park in Battersea, in association with the Festival of Britain, in 1951.[55] But for a county council, the equivalent tier of government, to devote public finance to establishing an amusement park of this kind would be unthinkable, even in the most expansive and self-confident phase of local government activity in the late nineteenth and early twentieth century. Country parks, emphasising nature, picnics, education and outdoor exercise, were a different matter. The nearest approximation might be the London County Council's programme of building and operating lidos between the two World Wars.[56] Outside socialist countries, however, amusement parks of the Artxanda type were almost invariably the preserve of private enterprise, in Britain and more generally.

The artist Guillermo Santamaría has suggested that the Artxanda amusement park was ahead of its time, and would have been a success thirty years on, when the cultures of consumerism, commercial leisure and the private car on which it depended had become much more firmly established, and in the light of Bilbao's postmodern development as a tourist destination in the wake of the highly successful Guggenheim museum project.[57] In support of this contention we might look to the opposite coast of northern Spain, to the highly successful Port Aventura theme park, which was promoted by the government of the autonomous community of Cataluña (a step up from a provincial Diputación) in conjunction with the local authorities. The planners assembled international capital (from the United States and Britain) as well as regional resources, in the face of several administrative and financial setbacks. Work began in 1992, only two years after the closure of Artxanda, but the park did not open until May 1995. But this was a new concept for Spain, self-consciously conceived as a Catalan response to Euro Disney; and it formed part of a planned regeneration programme for a mature (and troubled) international tourism destination. This commitment to attracting tourists from overseas, as opposed to providing an amenity for local people, marked out a sharp contrast between Artxanda and Port Aventura. It entailed investment on an altogether different scale (300 million euros at the outset, which makes Artxanda's investment – and losses – look tiny), and operated on a global stage.[58]

This comparison suggests that the *Parque de Atracciones de Vizcaya*, on its cramped, inaccessible, windswept, cloud-capped mountain site, was too small, too soon and too parochial, and too closely tied in to the political expedients of a dying regime and the collapse of an old industrial model, to stand any chance of long-term survival. This did not prevent it from earning the retrospective, nostalgic love and affection of a generation of *Bilbaínos*, and from inspiring the commemorative creativity of artists, bloggers and online archivists; but this celebration of lost heritage through virtual homage was unable to bring it back to life. As Joni Mitchell once put it, 'You don't know what you've got till it's gone'.[59] This was not Paradise, and it already had a parking lot, though the pink hotel of the song was absent; but this did not still the lamentations of those erstwhile (in many cases occasional) patrons who now learned to cherish their vivid but fugitive memories of the lost glories of Artxanda, and of their childhood. Many questions about the amusement park's genesis, intended purpose, management and demise remain unanswered, at least for the time being; but its trajectory, and especially its afterlife, demonstrate the power of the amusement park idea to command attachment to iconic place and remembered or imagined pleasure, an assertion of topophilia in defiance of the brute alternative realities of profit, subsidy and loss; here as in so many other locations across the globe.[60]

## Notes

1  Susana Serrano, Pedro A. Novo and José Maria Beascoechea, 'Un Siglo y Media del Bilbao Metropolitano: Una Visión Desde la Historia Urbana', in Manuel González Portilla, José Maria Beascoechea Gangoiti and Karmele Zarraga Sangroñiz (eds), *Procesos de Transición, Cambio e Innovación en la Ciudad Contemporánea* (Bilbao: Servicio Editorial de la Universidad del País Vasco, 2011), pp. 635–50.
2  I have adopted the Basque spelling 'Artxanda' throughout. The Spanish spelling 'Archanda' was official usage under the Franco regime.

3  http://esperandoaltren.blogspot.co.uk/2010/10/el-parque-de-atracciones-de-artxanda-1.html accessed 22 April 2013. This excellent website, which is remarkable for the quality of its photographic material, presents the Artxanda park in two parts, the second of which has a '-2' suffix. It is unpaginated and will be cited hereafter as 'Esperando al Tren: Artxanda', Part 1 or 2.
4  'Esperando al Tren: Artxanda', Part 2.
5  Luis Castells (ed.), *El Rumor de lo Cotidiano* (Bilbao: Universidad del País Vasco, 1999).
6  Nigel Townson (ed.), *Spain Transformed: The Late Franco Dictatorship* (Basingstoke: Palgrave Macmillan, 2007).
7  Rocío García Abad, Manuel González Portilla, Arantza Pareja Alonso and Karmele Zarraga Sangroniz, 'Migraciones Interiores en el Ciclo Industrial de la Ría de Bilbao (1876–1975)', in González Portilla, Beascoechea Gangoiti and Zarraga Sangroniz, *Procesos de Transición*, p. 216, Table 9.1.
8  José Luis de la Granja, Santiago de Pablo and Coro Rubio Pobes, *Breve Historia de Euskadi. De los Fueros a la Autonomía* (Madrid: Debate, 2011).
9  Manuela Aroca Mohedano, *El Sindicalismo Socialista en Euskadi (1947–1985)* (Madrid: Biblioteca Nueva, 2013); Fernando Molina, '*Fagor Electrodomésticos*: The Multinationalisation of a Basque Co-operative, 1955–2010', *Business History* 54 (2012), pp. 945–63; Fernando Molina and John K. Walton, 'An Alternative Co-operative Tradition: The Basque Co-operatives of Mondragón', in Anthony Webster, Alyson Brown, David Stewart, John K. Walton and Linda Shaw (eds), *The Hidden Alternative: Co-operative Values Past, Present and Future* (Manchester and Tokyo: Manchester University Press and United Nations University Press, 2011), pp. 226–50.
10  José Antonio Pérez, *Los Años del Acero* (Madrid: Biblioteca Nueva, 2001), ch. 5–6.
11  Rafael Aguirre Franco, 'Bases para una Política de Turismo en San Sebastián y su Zona', mimeographed policy document (San Sebastián, Centro de Atracción y Turismo, Junio 1971 – author's collection), pp. 4–6, 12, 31–2; John K. Walton, 'Another Face of "Mass Tourism": San Sebastián and Spanish Beach Resorts under Franco, 1936–1975', *Urban History* 40 (2013), pp. 483–506.
12  *ABC* (19 June 1974).
13  Ramiro Pinilla, *Guía Secreta de Vizcaya* (Madrid: Al-Borak, 1975), pp. 154–64.
14  Peter Bailey, *Leisure and Class in Victorian England* (London: Routledge and Kegan Paul, 1978).
15  Gaizka Fernández Soldevilla and Raúl López Romo, *Sangre, Votos, Manifestaciones: ETA y el Nacionalismo Vasco Radical, 1958–2011* (Madrid: Tecnos, 2012); Maria Victoria Gómez García, *La Metamorfosis de la Ciudad Industrial* (Madrid: Talasa, 2007).
16  http://www.tibidabo.cat/es/historia accessed 22 April 2013; Gary S. Cross and John K. Walton, *The Playful Crowd: Pleasure Places in the Twentieth Century* (New York: Columbia University Press, 2005), ch. 2–3.
17  John K. Walton, *Riding on Rainbows: Blackpool Pleasure Beach and its Place in British Popular Culture* (St Albans: Skelter, 2007), p. 14; Laurentino Gómez Beldarrain, *San Sebastián: Historia de los Parques de Recreo a Través de la Tarjeta Postal* (San Sebastián: Elkar, 2005); Juan José Olaizola Elordi, 'Ferrocarril y Turismo en San Sebastián (1864–1914)': http://www.docutren.com/congreso_vitoria/comunicaciones/3021.pdf accessed 6 May 2013, pp. 7–16.
18  http://www.parquedeatracciones.es/historia-del-parque accessed 22 April 2013.
19  Paloma Santamaría, 'Ún Paseo por Parques de Atracciones Abandonados', *ABC* (14 July 2012): http://www.abc.es/20120714/estilo-viajes-absi-parques-atracciones-abandonados-201207101411.html accessed 22 April 2013.
20  Walton, *Riding on Rainbows*, pp. 71–3; Cross and Walton, *The Playful Crowd*, ch. 3–4.
21  Susana Serrano Abad and José Maria Beascoechea Gangoiti, 'La Ría de Bilbao (1975–2000). Hacia un Modelo Metropolitano Post-industrial', in Carlos Contreras Cruz

and Claudia Patricia Pardo Hernández (eds), *La Modernización Urbana en México y España, Siglos XIX y XX* (Puebla: Benemérita Universidad Autónoma de Puebla, 2009), pp. 431–64.

22  Gómez García, *La Metamorfosis de la Ciudad Industrial*, pp. 118–35.
23  Luis Bilbao Larrondo, 'La Metamorfosis de Bilbao. 1975–1979', *Ondare* 26 (2008), pp. 287–300; Gómez García, *La Metamorfosis de la Ciudad Industrial*, pp. 136–40.
24  Luis C. Núñez Astrain, *La Sociedad Vasca Actual* (San Sebastián: Txertoa, 1977), pp. 15–40.
25  Eugenia Afinoguénova and Jaume Martí-Olivella (eds), *Spain is (Still) Different* (Lanham, MD: Lexington Books, 2008).
26  'Esperando al Tren: Artxanda', Part 1, provides this quotation. All translations from Spanish have been undertaken by the author, who takes responsibility for them.
27  Duncan Shaw, *Fútbol y Franquismo* (Madrid: Alianza, 1987); Ángel Bahamonde, *El Real Madrid en la Historia de España* (Madrid: Taurus, 2002), which provides a more balanced view of Real Madrid; John K. Walton, 'Basque Football Rivalries in the Twentieth Century: Real Sociedad and Athletic Bilbao', in Gary Armstrong (ed.), *Fear and Loathing: Local Rivalries in World Football* (Oxford: Berg, 2001), pp. 119–33.
28  'Esperando al Tren: Artxanda', Parts 1 and 2.
29  'Esperando al Tren: Artxanda', Part 1.
30  Frédéric Chaubin, *CCCP: Cosmic Communist Constructions Photographed* (Cologne: Taschen, 2011).
31  'Esperando al Tren: Artxanda', Part 1.
32  Raúl López Romo, *Euskadi en Duelo. La Central Nuclear de Lemoniz como Símbolo de la Transición Vasca*, and Mikel Alonso, *Lemóniz* (San Sebastián: Hariadna, 2012); Juan Luis Olaran Sustatxa, *El Contubernio Nuclear: Lemoiz* (Gasteiz: Arabera, 2010).
33  'Esperando al Tren: Artxanda', Part 2.
34  Imanol Villa, 'En el País de los Chimbos', *El Correo* (9 August 2009); Juan M. de Pertika, 'Los Tximbos', *Munibe* (1954), pp. 308–11: http://www.aranzadi-zientziak.org/fileadmin/docx/Munibe/1954308311.pdf accessed 12 May 2013.
35  Jon Juaristi, *El Chimbo Expiatorio (La Invención de la Tradición Bilbaína (1876–1939))* (Bilbao: El Tilo, 1994).
36  Mercedes Aguirre and Alicia Esteban, *Cuentos de la Mitología Vasca* (Madrid: De La Torre, 2006), pp. 92–6.
37  Xosé M. Núñez Seixas, *Los Nacionalismos en la España Contemporánea* (Barcelona: Hipòtesi, 1999), ch. 8 sets this process in context.
38  'Esperando al Tren: Artxanda', Part 2; Oskar L. Belategui, 'Las Reliquias de Aquel Parque de Artxanda', *El Correo* (13 June 2012).
39  'Esperando al Tren: Artxanda', Part 1.
40  Pinilla, *Guía Secreta de Vizcaya*, p. 144.
41  http://www.bilbao.net/cs/Satellite/funicularArtxanda/Breve-historia-del-Funicular-de-Artxanda/es/1000001146/Contenido accessed 22 April 2013; *ABC* (7 July 1971; 16 February 1975).
42  *ABC* (22 June 1974; 23 July 1975, 2 October 1975).
43  'Esperando al Tren: Artxanda', Part 1; Belategui, 'Las Reliquías'.
44  Cross and Walton, *The Playful Crowd*, ch. 6.
45  *ABC* (5 June 1974).
46  Cross and Walton, *The Playful Crowd*, p. 207.
47  'Esperando al Tren: Artxanda', Part 1.
48  'Esperando al Tren: Artxanda', Part 2.
49  'Esperando al Tren: Artxanda', Part 2.
50  Carlos Humanes, 'Bankunión fue Incapaz de Superar la Crisis Económica y Adaptarse a las Nuevas Estructuras Bancarias', *El País* (11 April 1982).
51  'Esperando al Tren: Artxanda', Part 2; Belategui, 'Las reliquías'.
52  Belategui, 'Las reliquías'.

53   http://www.vuelvenlasatracciones.com accessed 22 May 2013; Saioa Olmo, *Vuelven las Atracciones* (Bilbao: Consonni, 2010). The names of the provincial authorities had changed after the advent of Basque autonomy in 1982.

54   https://es.wikipedia.org/wiki/Parque_de_Atracciones_de_Vizcaya accessed 23 August 2013.

55   Becky E. Conekin, *'The Autobiography of a Nation': The 1951 Festival of Britain* (Manchester: Manchester University Press, 2003); also see Ian Trowell, this volume.

56   Janet Smith, *Liquid Assets: The Lidos and Open Air Swimming Pools of Britain* (London: English Heritage, 2005).

57   Belategui, 'Las reliquías'.

58   S. Anton Clavé (ed.), *Lecciones Sobre Turismo: El Reto de Reinventar los Destinos* (Barcelona: Planeta, 2012), pp. 170–72.

59   Joni Mitchell, *Big Yellow Taxi* (1970), http://jonimitchell.com/music/song.cfm?id=13 accessed 23 May 2013.

60   Yi-Fu Tuan, *Topophilia* (New York: Columbia University Press, 1974); John Bale, *Sport, Space and the City* (London: Routledge, 1993).

# Part III

# Cultural significance, revival and heritage protection

# 9 Last night of the fair

## Heritage, resort identity and the closure of Southport's Pleasureland

*Anya Chapman and Duncan Light*

The seaside resort of Southport in north-west England, 17 miles north of Liverpool, has long been preoccupied with issues of social tone. In particular, the town has always imagined itself as being a select and refined place and has never enthusiastically embraced mass tourism nor its associated popular entertainments. Consequently, the place has long had an uneasy relationship with its amusement park, Pleasureland. Yet, from its opening in 1922, Pleasureland became a well-established landmark within Southport. It was universally known, widely visited, despised by some, but held in considerable affection by others. Indeed, many people in Southport regarded Pleasureland as a part of the resort's heritage. However, this was an 'unofficial heritage'.[1] It did not enjoy any official recognition or formal protection but it was a site that had meaning and significance for (some of) the local community and it was important within the personal biographies and memories of many in the town and surrounding region.

This chapter will explore Southport's ambivalent relationship with Pleasureland by examining the early growth of the amusement park, its establishment as a local landmark and the circumstances of its sudden closure in 2006. While many in the town were deeply saddened at the closure of the amusement park, the local authority regarded this as an opportunity to redevelop the site in accordance with a broader project to rebrand Southport as a refined and sophisticated 'Classic Resort'. Ultimately, these plans came to nothing so that a new funfair was established on the former Pleasureland site. But the local authority still aspires to a landmark development of the site and regards an amusement park as incompatible with plans to gentrify the resort.

## The rise and fall of Southport's Pleasureland

Southport is one of England's earliest coastal resorts, having started to grow in the late eighteenth century. From the outset, the resort was intended to be a select place for the affluent to live, take holidays and retire to. Since two families owned most of the land in and around the town they were able to control and regulate the nature of the resort's development.[2] As such the town was carefully laid out with wide tree-lined boulevards, parks and gardens, and large houses and villas. The arrival of the railways, however, thwarted the resort's aspirations towards refinement and gentility. In 1850, the town was connected by rail to Liverpool; a direct line to Manchester opened in 1862; a line to Preston opened in 1882; and in 1884 a second line arrived from Liverpool. Southport was now easily accessible from the neighbouring cities (particularly Liverpool and Manchester) and consequently the number of visitors to the town increased rapidly. In 1866 alone, 800,000 people arrived in the resort by train.[3] Nevertheless, Southport continued to imagine itself as being a more refined type of resort in which mass tourism was tolerated rather than encouraged.

Given the type of resort that its landowners wished to create, there was, unsurprisingly, an initial resistance to catering for popular amusements for visitors.[4] Nevertheless, an informal collection of sideshows and stalls gathered on the promenade near to the pier. The retreat of the sea in the second half of the nineteenth century gave the town corporation the opportunity to remodel the resort using reclaimed land on the foreshore. This area was allocated for public recreation and the provision of facilities that were appropriate to Southport's aspirations towards gentility. A Marine Lake was created between 1887 and 1895, along with a new road (Marine Drive), which established a new promenade almost half a mile seaward of the original promenade (1835–38). The sideshows were pushed to a new site on reclaimed land at the southern end of the town and, by the 1890s, a ramshackle and haphazard fairground had become established (Figure 9.1).[5] This fairground was a popular attraction for visitors to the town and quickly gained a reputation for the disorderly behaviour and spirit of the carnivalesque, which characterised seaside holidays.[6] The presence of the fairground – and particularly its proximity to Lord Street (the town's refined central boulevard), the Winter Gardens and one of the town's railway stations – caused some disquiet among the town's residents and middle-class visitors. However, while 'those with political power and influence wanted to keep the town as genteel as possible, there was a conflict of interest with those catering for the demands of the holiday trade'.[7] As a result, the local council allowed the fairground to remain. It was renamed 'The White City' in 1911, seemingly in an attempt to give it a form of respectability, and appears to have been accepted by the town's residents.[8]

This uneasy truce continued until the end of 1912, when the council decided to undertake further improvements to Southport's foreshore and promenade. Consequently, the White City was moved to a new site further seaward. Stephenson's *Guide to Southport*, published in 1913, indicated how successful the landscaping and 'improvement' of the amusement park had been: 'former visitors to Southport will remember the old Fair Ground, which consisted of dilapidated wooden shanties, Aunt Sallies, coconut shies and so forth. All this rubbish has been cleared away, and what is fittingly termed 'The White City' put in its place'.[9] The White City was closed during the First World War and, in 1922, was moved again to a new site on reclaimed land by the new promenade. It reopened under a new name: Pleasureland (Figure 9.2).

The new Pleasureland site indicates the continued involvement of the local authority in the planning of Southport's development in a way that was unlike many other resorts. In particular, the influence of modernist urban planning – particularly

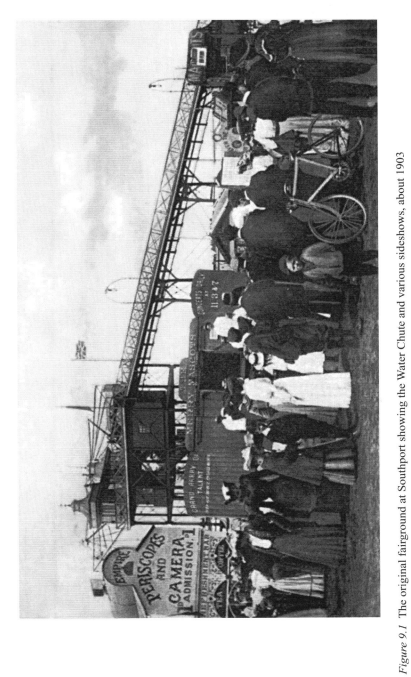

*Figure 9.1* The original fairground at Southport showing the Water Chute and various sideshows, about 1903

Source: © www.facebook.com/Southport-Yesteryear

*Figure 9.2* A view from the top of the relocated Water Chute showing the newly established Pleasureland, about 1924

Source: © www.facebook.com/Southport-Yesteryear

the application of land use zoning – is apparent. The new Pleasureland site was larger than that of The White City but it was also spatially set apart from the rest of the town, and particularly the original promenade and Lord Street. The site was clearly bounded, by the seafront to the west, a railway line and sand dunes to the south, the Marine Lake and Kings Gardens to the east, and Princes Park to the north. It was also partially enclosed by a white wall on the landward side in which an elegant clock-tower gate was added in 1926. Pleasureland was now effectively a mono-functional enclavic space[10] where carnivalesque behaviour was carefully contained so as not to disrupt the social tone of the resort as a whole. The local authority had devised a compromise to permit and regulate the town's amusement park in a way that was acceptable to the town's residents and appropriate to the ambience and image of the wider resort.

Pleasureland proved to be a very successful amusement park. During the 1920s and 1930s the park expanded and new rides were added including, in 1937, the Cyclone rollercoaster (Figure 9.3). It was also profitable for the local authority.[11] The site was mothballed for most of the Second World War but Pleasureland subsequently reopened and regained some of its former popularity during the 1950s and 1960s. However, when British seaside resorts began to lose patronage from the late 1960s onwards, Southport was badly affected since, unlike Blackpool, the resort did not have a loyal, returning working-class market to depend upon. By the 1970s, Pleasureland was badly in need of investment, which the local authority was now unable to provide. Moreover, the local government reorganisation of 1974 had seen the county borough of Southport abolished and absorbed into Sefton Borough Council, which covered a large part of the northern suburbs of Liverpool. For various reasons the new council was unable to invest in Pleasureland. Consequently, in 1981, Blackpool Pleasure Beach leased the site (which included many tenanted rides and attractions) from Sefton Council. The new owners undertook to develop the park and introduced a number of new rides. However, since almost half of the attractions and rides were still owned by tenants, it took the Pleasure Beach company until 2004 to acquire the entire site.

During the late 1990s, Southport was involved in numerous projects intended to revive and regenerate the resort. These included building new hotels, conference facilities, an indoor water park, the refurbishment of the pier, and the building of a new sea wall and associated retail and leisure park. In this context Blackpool Pleasure Beach undertook a major programme of investment in Pleasureland. A new brand, 'Amusement Park on the Sands', was created[12] and the park underwent a process of landscaping and theming from 1997. In particular, there was an attempt to transform Pleasureland into a form of theme park. The theming was loosely based on Morocco so that Pleasureland now featured attractions such as the Casablanca Entertainment Centre and rides such as The Lost Dinosaurs of the Sahara, Abdullah's Dilemma, The Marrakech Express, The Desert Rescue and King Solomon's Mines. The park also introduced a mascot by the name of Ali-Ba Bear. This was a late twentieth-century example of the Orientalism that had long pervaded seaside architecture.[13] Not everything, however, was themed and in

*Figure 9.3* General view of Pleasureland showing the Cyclone rollercoaster, 1970s

Source: © www.facebook.com/Southport-Yesteryear

1999 a suspended, looping rollercoaster called The Traumatizer was installed at a cost of £5 million.

These efforts to rebrand Pleasureland as a theme park were only partially successful. The physical fabric of the site had certainly been transformed through landscaping, the themed environment, and through branded merchandise, catering outlets and rides. Despite this, the long-established carnivalesque culture of the fairground persisted. Pleasureland was a place for unrestrained enjoyment. Like the joyful but disorderly and sometimes violent crowd that characterised Blackpool Pleasure Beach in the post-war decades,[14] incidents of violence and aggression amongst customers and staff continued to occur at Pleasureland. Similarly, intoxication, illicit sexual encounters and the use of lewd or inappropriate behaviour and language was commonplace among both customers and employees. There was, then, a clash of cultures between the sanitised, commodified and regulated theme park, and the disorder, brashness and elation of the seaside fairground. Not for nothing did many local residents (and the park's regular customers) continue to refer to the attraction as 'the fair'.

Moreover, throughout its history Pleasureland had been an open park. It was, therefore, a 'democratic, accessible space'[15] that was easy to enter. As a result, Pleasureland was used in a wide range of ways by a broad range of people, many of whom were not customers. For example, local people would wander through the park to walk their dogs. Children gathered there (particularly during school holidays) not to use the rides but to make mischief. Young people used the park as a place to meet, engage in under-age drinking (with alcohol that they had brought with them) or take recreational drugs. Others went there intent on theft (through breaking into machines) or in search of a fight, particularly on so-called 'scally Sundays'. And some went just to gaze upon everything else that was taking place. In other words, many of Pleasureland's consumers used the site for a wide range of recreational activities that generated no revenue for the park. Even some of the 'legitimate' customers went there with the intention of spending as little as possible.

Pleasureland was a very popular attraction, attracting more than 2 million visitors, and was consistently one of the top five free attractions in England. Its only competitor in the theme park market was Blackpool Pleasure Beach. Further investments totalling £2 million were made at Pleasureland in 2004.[16] However, in June 2004, Geoffrey Thompson, the managing director of both Blackpool Pleasure Beach and Pleasureland, died suddenly and his daughter took over both companies. Local fears that she did not share her father's commitment to Pleasureland appeared to be borne out when in September 2004 Pleasureland introduced a £2 entry charge, effectively becoming a 'closed' park. As Table 9.1 shows there was a dramatic drop in visitor numbers the following year.

On 5 September 2006, Pleasureland suddenly closed. There had been no prior indication that closure was imminent. A press statement[17] claimed that the park had proved unsustainable with none of the recent investments generating a return on capital. It also pointed to competition from publicly funded and lottery-funded attractions and from Sunday shopping and sporting events. In the following months the park's assets were removed or demolished and the speed at which this

*Table 9.1* Visitor numbers at Pleasureland and Blackpool Pleasure Beach[18]

| Year | Pleasureland | Blackpool Pleasure Beach |
|------|--------------|--------------------------|
| 1991 | 2.0 million* | 6.5 million* |
| 1992 | 2.0 million* | 6.5 million* |
| 1993 | 2.0 million* | 6.75 million* |
| 1994 | 2.0 million* | 6.75 million* |
| 1995 | 2.0 million* | 7.3 million* |
| 1996 | 2.0 million* | 7.5 million* |
| 1997 | 2.1 million* | 7.8 million* |
| 1998 | 2.1 million* | 7.1 million* |
| 1999 | 2.5 million* | 7.2 million* |
| 2000 | 2.6 million* | 6.8 million |
| 2001 | 2.0 million* | 6.5 million |
| 2002 | 2.0 million* | 6.2 million |
| 2003 | 2.1 million* | 6.2 million |
| 2004 | 2.1 million* | 6.2 million |
| 2005 | 0.5 million | 6.0 million |
| 2006 | No data | 5.73 million |
| 2007 | – | 5.5 million |

* Estimate

was organised led many Southport residents to speculate that the closure had been planned for some time (Figure 9.4). Some of the more popular rides were moved to Blackpool Pleasure Beach, while others were sold to other theme parks. Some of the original rides were donated to the proposed Dreamland heritage amusement park in Margate (Figures 9.5 and 9.6).[19] However, in a move that came as a considerable shock to local people, the Cyclone rollercoaster was demolished in November 2006 (Figures 9.7 and 9.8). There were calls by Southport residents for intervention by the local authority. However, while the Cyclone had previously been considered for listing by English Heritage, its unlisted status meant that there was no breach of planning regulations and the local authority was unable to take any action.[20] By all accounts, the site was left in a near derelict condition, and local people speculated that the former owners had sought to ensure that it could not be used as a fairground or amusement park by a rival operator. As a result, Sefton Council needed to spend £350,000 to make the site safe.[21]

## Reactions to Pleasureland's closure – The local community

An examination of forum postings on a popular local website (www.southportgb. com) reveals the local community's response. Obviously, like any Internet forum, these have to be used with care since they only divulge the views of forum users and do not constitute a strictly representative cross-section of local opinion. Nevertheless, they can be used to give a broad indication of how the town reacted to the demise of Pleasureland.[22]

*Figure 9.4* Pleasureland lies abandoned following the removal or demolition of a number of rides, 2007

Source: © Nick Laister

*Figure 9.5* The abandoned River Caves, 2007

Source: © Nick Laister

*Figure 9.6* The River Caves' boats ready to be relocated to Dreamland, Margate, January 2007

Source: © Anya Chapman

*Figure 9.7* The Cyclone rollercoaster cars before their removal to Blackpool Pleasure Beach, January 2007

Source: © Anya Chapman

*Figure 9.8* Part of the demolished Cyclone rollercoaster track, January 2007

Source: © Anya Chapman

Many of the initial reactions were of shock or sorrow. For example, one poster stated: 'It felt as if my guts had been ripped out when I heard this news. I just can't believe it'. Another stated: 'I am totally devastated at this news – and it seems that the majority of people share the despair I'm feeling'. Similarly: 'If somebody had told me that Pleasureland was closing I'd have thought it was an April Fool. But it's happened! It's just amazing! What a shame, I am absolutely gutted'. Some reactions were even stronger: 'I'm not ashamed to say that I actually cried. Maybe it's an over-reaction to some but it's my childhood being taken apart, something which said Southport to me . . . What's there here to be proud of any more?' These responses indicate that at least a section of the Southport population cherished Pleasureland and they point to the distress that some felt on hearing of its closure. However, such responses were not universal. One forum user stated dismissively: 'There's more to Southport than Pleasureland. We should be more positive and focus on what Southport has got and what it can offer in the future'. Another said simply: 'I did like it but it was always a bit tacky'. Another contributor effectively summed up Southport's ambivalence towards Pleasureland by stating: 'People want theme parks not fairgrounds. People have been complaining about Pleasureland for years, saying it's over-priced and run down. Those people got their wish on September 5th'.

Many users of the forum stressed their happy memories of visits to Pleasureland, particularly as children. One user stated simply: 'All my wonderful childhood memories, all gone in the blink of an eye'. Another recalled their childhood experiences:

> I used to love going to the fairground. My parents didn't have the money for us to go on many of the rides but we had great fun walking around and pestering people for the odd 10 pence so that we could have a go at something or other. It feels like that's the end of my happy childhood memories.

Another reminisced:

> I'm devastated. It really was part of my childhood. I used to come to Southport many times when I was a child. My parents preferred it to Blackpool. After a day on the beach the highlight of the day was a trip to Pleasureland for a few hours – so long as we'd behaved ourselves! And I've taken my kids to the fair many times.

In this context, Pleasureland can be identified as a material reference point within personal biographies. It is a site that anchored memories of living in – or visiting – Southport and for this reason many felt a strong emotional attachment to the park.

A number of the forum posts are of interest for the way in which they use the term 'heritage' to speak of Pleasureland. Until the relatively recent work by English Heritage (now Historic England) and others,[23] there has been a long-standing reluctance in Britain to consider the architecture and structures of seaside resorts in terms of heritage. Underpinning this approach is a set of ideas that

Laurajane Smith[24] has termed the 'authorised heritage discourse'. This attributes value to that which is material (and often monumental), aesthetically pleasing, authentic, associated with social elites, and significant in some way for national history and identity. Moreover, this discourse privileges the roles of experts and professional knowledge in defining and regulating the way in which societies think about heritage. Such a discourse prioritises particular understandings of what heritage is and subdues or marginalises alternative constructions of heritage.[25] As a result, the historic buildings of seaside resorts have long been ignored by 'official' heritage institutions. Such buildings were little valued in terms of age or aesthetics and, as sites associated with mass, popular entertainment, have been dismissed as banal or insignificant.

But while Britain's heritage institutions and agencies may have been indifferent to the heritage of fairgrounds and amusement parks,[26] many of Southport's residents explicitly considered Pleasureland in such terms. For example, one forum user stated directly: 'They don't have any right to destroy our heritage'. Another poster felt even more strongly:

> I feel like I want to padlock myself to it and go on a hunger strike. Let's all start a protest at this. The people of Southport saved the pier – why can't we do the same for Pleasureland. We shouldn't accept this lying down. It's a big part of our heritage.

Another argued: 'It's a monument that's too precious . . . for it just to be disregarded' and here the use of the term 'monument' gives Pleasureland an equal status with other, more traditional forms of heritage. There was particular outrage at the demolition of the Cyclone rollercoaster. In particular, a number of posters identified it as a local landmark that in some way symbolised the town itself. One argued: 'Cyclone has been a dominant part of the Southport skyline for almost 70 years. It will be sorely missed'. Another person was more explicit about the Cyclone as a local symbol:

> Heaven knows I am hardly sentimental about the town I grew up in but I could never imagine that the Cyclone – of all things – would go. More than any other landmark or building in the town it says 'Southport' to me.

Similarly: 'I really understand the pain that people feel having seen the Cyclone ripped apart like that. It was Southport. And it had a place in our hearts'.

Pleasureland – and the Cyclone in particular – illustrate how historic structures can be important for grounding senses of place and identity at the local scale.[27] Such structures can become common points of reference and identification around which community cohesion is built.[28] They can also be landmarks that are significant and meaningful in symbolising a community itself. Pleasureland, then, can be identified as a form of 'unofficial' heritage,[29] something that does not receive recognition by the state and other authorising institutions of heritage (such as the National Trust) but which, nevertheless, is important within local-level and popular relationships with the past. A similar example of a local, unofficial heritage in a seaside resort is the case of the herring-fishing industry in Great Yarmouth. Sheila Watson[30] notes that this industry was regarded as a relatively unimportant part of the town's economy by historians but nevertheless was deeply rooted in the collective memory of the town and was identified by local people as a key element of the town's heritage that they wished to see interpreted in a local museum. Yet, all heritages are contested and dissonant[31] and in this context the value attached to Pleasureland (and the Cyclone) was not universally shared. For example, one forum post sought to put things into perspective: 'It [the Cyclone] is a bloody roller-coaster, not Westminster Abbey'. Another poster was dismissive: 'It's a dilapidated old eyesore that should be knocked down'.

A number of forum posts took a broader perspective, fretting over the implications of the closure of Pleasureland for the future of Southport itself. In particular, some people worried that the town could not survive without a high-profile attraction such as an amusement park. This view was summed up by one post: 'Now that Pleasureland's closed what does Southport have left to offer visitors? A coastal resort that doesn't have an amusement park? It's disgraceful'. Similarly: 'What sort of a resort is it with no sea and no fairground? I'm shocked at the news today'. Others were concerned that Pleasureland was a landmark that had given Southport something different. One argued that: 'Without Pleasureland we're becoming like every other . . . seaside town', while another questioned: 'What reason have families got to come Southport now? Nothing'. But again, not all views were negative. Some comments argued that Pleasureland could not have been an appropriate anchor for a modern resort. One user stated:

> What we need are up-to-date facilities that will encourage people to come here for conferences and high-quality events. They would create more jobs than were lost at Pleasureland, they'd also bring money into Southport. We should be more positive. When times change we should adapt to the changing market.

There were undoubtedly many people in Southport who did not mourn the closure of Pleasureland. Nevertheless, the posts on the local Internet forum do indicate that Pleasureland (and the Cyclone) were regarded with considerable affection by a part of the Southport population. In addition, there was recognition of their importance for the town, both as visual landmarks, and as symbols of local heritage and identity. More broadly, there was also recognition of the economic importance of Pleasureland and its role as a 'marker'[32] of a seaside resort. Despite the local popularity of the park, however, subsequent events suggested that the local authority placed much less value on Pleasureland and regarded its closure as an opportunity (rather than a threat) for the town.

### Reactions to Pleasureland's closure – The local authority

In order to understand Sefton Council's response to the closure of Pleasureland, it is necessary to consider the broader context of regeneration initiatives in

Southport. In 2003, the Northwest Development Agency (a government-funded body with responsibility for promoting economic growth and regeneration in north-west England) published a report entitled *A New Vision for Northwest Coastal Resorts*.[33] This focused on a long-term strategy for the regeneration of resorts in the region. The report argued that tourism was unlikely to be the future driver of the local economy in such places (with the possible exception of Blackpool). Instead, the priority should be to transform these resorts into attractive places in which to live and work which would, in turn, attract visitors. The report was blunt about the need to remove 'the obsolete and unsightly trappings of the old tourism'[34] but without destroying the sense of place of these resorts.

The report produced visions for a number of resorts in north-west England that could form the basis for rebranding and regeneration strategies. For example, Morecambe's vision was 'beautiful place'; Fleetwood's was 'UK Capital of Value' and Blackpool's was 'Europe's No.1 Resort'. However, the report noted that Southport at that time had an unclear brand (due to the contrast between the elegance of Lord Street and the 'bucket and spade' offer of the seafront). It therefore proposed a new vision centred on 'style and sophistication'.[35] The report noted that Southport was a relatively prosperous local/regional centre and retained a reputation as a genteel place. It proposed to build on these strengths by developing a brand based around upmarket shopping, centred on an improved Lord Street.

The report also proposed the hallmark of the 'Classic Resort'.[36] This was to be a clear and recognisable brand that, above all, was characterised by quality and 'quiet sophistication'.[37] It argued that such places would preserve a sense of the traditional seaside holiday but attached to high quality for high-value consumers. It identified five core components of a Classic Resort: a promenade and beach with sea views; traditional attractions (including funfairs and parks); a range of accommodation for visitors; wet weather facilities for entertainment and relaxing; and leisure swimming areas (including open air bathing). It also proposed that Classic Resorts would be characterised by respect for the historic and environmental heritage of the resort; a high-quality built and natural environment; quality shopping; quality hotels; quality food and beverage; and a programme of cultural activities. What is significant about the *New Vision for Northwest Coastal Resorts* report is that it envisioned a place for Pleasureland in projects to rebrand Southport as a Classic Resort. A SWOT analysis of Southport identified the amusement park as one of the town's strengths (particularly since traditional seaside entertainment was identified as a core component of the Classic Resort). In this sense, the report explicitly identified Pleasureland as a part of the resort's heritage. It suggested that the park could be improved and extended but also argued that Pleasureland should reposition itself in order to target a higher-value audience (in a similar way to Alton Towers).[38]

Sefton Council was clearly attracted to the idea of the Classic Resort. After consulting with local stakeholders it commissioned a report (from the same consultants who had authored *A New Vision for Northwest Seaside Resorts*), to define what Classic Resort status would mean for Southport. The report[39] reiterated that

Southport had an unclear brand and it recommended that the town should abandon the remaining trappings of 'bucket and spade' tourism and develop a coherent destination brand image based on quality that would appeal to an affluent audience. This would enable Southport to define itself as *the* Classic Resort. The architectural heritage of the Victorian seaside resort was identified as an integral component of this new brand and the report argued that 'Southport's past is vital to its future as a Classic Resort'.[40] While the principal central shopping area (Lord Street) was central to such a brand, some regeneration and smartening of the seafront was necessary. Once again, Pleasureland was identified as one of the resort's strengths but the report called for extension and improvement to the amusement park.

It is uncertain how Blackpool Pleasure Beach (the owners of Pleasureland) reacted to these reports. The Classic Resort vision for Southport required major investments in Pleasureland, which its owners probably could not afford. In any case, an up-market theme park was not, at that time, a part of Blackpool Pleasure Beach's strategy. On the other hand, the introduction of an admission charge in 2005 may have been a first attempt to rebrand Pleasureland (and might have been intended to generate revenue to enable this to happen). The admission charge almost immediately changed the visitor profile of the theme park and had the effect of excluding the types of visitor (and associated ways of behaving) that were inconsistent with the vision of a new, gentrified Pleasureland.

In 2006, the Southport Partnership produced a pamphlet[41] intended for the residents of Southport, which explained the Classic Resort concept and set out how Southport was intending to achieve it. The document reiterated the importance of preserving the spirit of the traditional English coastal resort, with references to Southport's 'elegant seafront' and resources such as the pier, Marine Lake and gardens. However, Pleasureland (and indeed any reference to an amusement park on the seafront) was now conspicuously absent from the future vision for the town. While the tradition of popular entertainment was regarded by many as part of Southport's heritage, this was a heritage that the local authority appeared to consider as incompatible with the Classic Resort concept. Moreover, the intangible heritage of the fairground – the anarchic and disorderly behaviour characteristic of the carnivalesque – was actively dissonant with the quiet sophistication envisaged of a Classic Resort.

Following the closure of Pleasureland in September 2006, and the subsequent removal or demolition of the rides, the abandoned 25-acre site was an unwanted eyesore in an aspiring Classic Resort. Sefton Council re-acquired the site in April 2007 by purchasing the lease from Blackpool Pleasure Beach for £7.25 million.[42] This created an opportunity for a major re-development of the site in a way that accorded with the Classic Resort vision and which would complement other planned investments in the town.[43] Perversely, in preference to leaving the site empty in the short term, Sefton Council issued an 18-month licence to the Dreamstorm company to operate a fairground and consequently 'New Pleasureland' opened in July 2007. Dreamstorm were apparently eager to invest in redeveloping the site as a major amusement park. The local authority, however, had other plans that they regarded as more in keeping with an aspiring Classic Resort.

In March 2008, Sefton Council announced that Urban Splash (a Manchester-based company with a distinguished record in regenerating industrial and Victorian buildings) had been chosen to redevelop the Pleasureland site. The proposed 'Southport Marine Park' would include a landmark atrium, which would house a winter garden (inspired by the Eden Project in Cornwall), heated outdoor pool, new hotels and visitor accommodation, and an expanded Marine Lake. The £80 million investment was described as a 'reinvention of the seaside resort'[44] that would transform the recreational opportunities of the waterfront zone as well as creating an attraction of regional significance. Such a high-quality, eco-friendly attraction was entirely in keeping with the vision of a Classic Resort but also marked a decisive rejection of the fairground (and its associated culture) that had sat uneasily with Southport's new aspirations. The proposal was met with mixed opinions in Southport: many welcomed it, but others regretted that the funfair would not be able to stay on the site.

In the event Southport Marine Park was never built as funding for the project could not be secured because of the global recession of 2008–09. The project was formally abandoned in early 2012.[45] A spokesman for Sefton Council stated that the council remained committed to exploring possibilities for a long-term redevelopment of the site but had no funding to make any investments itself.[46] New Pleasureland had remained on the site during the 2008–12 period on a series of short-term leases. The owner (Norman Wallis) was able to introduce new rides and attractions and carried out environmental improvements but the uncertainty about the future of the site deterred major investments. But once the Southport Marine Park project was abandoned, Sefton Council offered Wallis a five-year lease (with talk of future rolling contracts). Certainly, in the medium term, there is little likelihood of any major development on the New Pleasureland site, so that Southport will continue to have a funfair rather than a theme park on the seafront (Figure 9.9).

While the local authority appears to regard New Pleasureland as 'better than nothing', the debates on the SouthportGB forum indicate that local people are supportive of it. One poster stated:

> OK so it's not great, it's not Alton Towers or Blackpool Pleasure Beach. But it's not bad either. You can see that a lot of effort and work has been made to improve it from the wasteland that it used to be.

Some clearly feel that a funfair or amusement park is crucial to the town's identity and economy as a seaside resort. One poster stated: 'Southport must have a fairground. People should get behind it . . . it can do nothing but good for our town', while another argued that: 'Southport needs a family-orientated funfair'. There also appears to be considerable local support for the funfair's owner who is regarded as an entrepreneur who is prepared to invest in the future of New Pleasureland (in contrast to the previous owners). One forum post states: 'I'm genuinely delighted that someone has worked tirelessly to make something new out of the mess left by the previous fairground owners'. Another argued: 'It's been

*Figure 9.9* Part of the New Pleasureland site, April 2015

Source: © Jason Wood

a difficult job for Norman Wallis but over the past 3 years he has turned the place around . . . it was fantastic to see so many visitors having fun at the fairground on a glorious Bank Holiday'. Wallis himself has continued to argue that Southport needs an amusement park.[47] In particular, he has argued that such a resource is entirely consistent with Southport's Classic Resort status, claiming 'what is a classic resort without rides and attractions?' He also argues that it brings a type of visitor – families – that are appropriate for a Classic Resort.

Nevertheless, an amusement park does not feature in the local authority's long-term vision for Southport. Sefton Council's draft local plan for the 2015–30 period[48] continues to allocate the New Pleasureland site to a major new tourism development that will be taken forward as a public–private partnership as circumstances permit. In particular, the plan talks of 'a significant opportunity to deliver high-quality development of a scale that enhances Southport's role as a regional tourism destination'. There is also an emphasis on quality design and high-quality landscaping of the new development. An amusement park is firmly excluded from Southport's push to achieve Classic Resort status and New Pleasureland is unlikely to have a long-term future on its current site.

## Conclusion

The case of Pleasureland illustrates differing conceptions of heritage in a seaside resort. For local planners and policy-makers, Pleasureland represents an unwanted heritage. The amusement park is associated with a form of holiday-making – the traditional 'bucket and spade' seaside holiday – which is now regarded as obsolete. It is also associated with a type of visitor – mass tourists – that has long been at odds with Southport's self-image (and is certainly discordant with the aspirations to gentrify the town). And the spirit of the carnival which characterises the amusement park – a part of the intangible heritage of the fairground – is firmly at odds with the 'quiet sophistication' envisaged of a Classic Resort. As a resource catering for mass popular entertainment, Pleasureland did not fit with the local authority's aspirations to rebrand Southport.

There were, and are, however, both internal and external audiences for Pleasureland. Visitors from outside the town – the external audience – accounted for the majority of Pleasureland's customers (and were probably responsible for most of the disorderly behaviour that took place there). But Pleasureland also had an 'internal' audience – the people of Southport. While many local residents were indifferent to the amusement park and rarely, if ever, visited it, many others in the town had strong emotional connections with Pleasureland, grounded in their own nostalgic memories of visiting. Pleasureland was a valued local landmark and an emblem of Southport's traditions as a seaside resort. For many it was a form of unofficial, popular heritage that was important for grounding local senses of place, past and community.

A key question, then, is: can there be a place for an amusement park in a Classic Resort? Put another way, is an amusement park incompatible in principle with the brand of a Classic Resort? On one hand, an amusement park is a well-established part of the heritage of the English seaside resort and is, therefore, entirely in keeping with a Classic Resort (as the current operator of New Pleasureland has argued). But on the other hand, an amusement park is not regarded as the type of high-quality product for a resort seeking to rebrand around style and sophistication. A more acceptable alternative might be a theme park – a 'destination park' in its own right – targeted at a high-quality family audience (like Alton Towers or Thorpe Park). Certainly the original Northwest Development Agency report that proposed the Classic Resort hallmark brand envisaged a place for an improved 'theme park' in Southport. But this is unlikely to be a realistic prospect.[49] The market for destination parks is uncertain and given the necessary investment it is unlikely that any large entertainment company would take the risk of building one in Southport. In short, then, the town's choice is between a traditional amusement park (like New Pleasureland) or something entirely different, and the local authority has chosen the latter option. Many people – both Southport residents and visitors – will regret that, as the town achieves Classic Resort status in coming years, an iconic part of the heritage of the seaside resort will be notable by its absence.

## Notes

1 Rodney Harrison, 'What is Heritage?' in Rodney Harrison (ed.), *Understanding the Politics of Heritage* (Manchester: Manchester University Press/Open University, 2010), pp. 8–9; Rodney Harrison, *Heritage: Critical Approaches* (London: Routledge, 2013), p. 15.
2 John K. Walton, *Wonderlands by the Waves: A History of the Seaside Resorts of Lancashire* (Preston: Lancashire County Books, 1992), p. 33.
3 Frank Bamford, *Back to the Sea: The True Story of Southport* (Southport: Frank Bamford, 2001), p. 29.
4 John K. Walton, 'The Demand for Working-class Seaside Holidays in Victorian England', *Economic History Review* 34, 2 (1981), p. 251.
5 Stephen Copnall, *Pleasureland Memories: A History of Southport's Amusement Park* (St Albans: Skelter Publishing, 2005), p. 11.
6 See for example: Rob Shields, *Places on the Margin: Alternative Geographies of Modernity* (London: Routledge, 1991), pp. 93–4; John K. Walton, *The British Seaside: Holidays and Resorts in the Twentieth Century* (Manchester: Manchester University Press, 2000), pp. 18–19.
7 Copnall, *Pleasureland Memories*, p. 13.
8 Ibid., pp. 21, 22.
9 Cited in Ibid., p. 21.
10 Tim Edensor, *Tourists at the Taj: Performance and Meaning at a Symbolic Site* (London: Routledge, 1998), pp. 45–6.
11 Copnall, *Pleasureland Memories*, p. 39.
12 Ibid., p. 81.
13 Fred Gray, *Designing the Seaside: Architecture, Society and Nature* (London: Reaktion, 2006), pp. 91–106.
14 John K. Walton, *Riding on Rainbows: Blackpool Pleasure Beach and its Place in British Popular Culture* (St Albans: Skelter Publishing, 2007), pp. 134–5.
15 Walton, *Riding on Rainbows*, p. 121.
16 BBC News, 'Historic Amusement Park to Close' (5 September 2006): http://news.bbc.co.uk/1/hi/england/merseyside/5318344.stm accessed 16 November 2013.
17 http://www.southportreporter.com/256/pleasureland-press-release.bmp accessed on 16 November 2013.
18 Max Hanna, *Sightseeing in the UK 1991* (London: British Tourist Authority/English Tourist Board Research Services, 1992); *Sightseeing in the UK 1992* (London: BTA/ETBRS, 1993); *Visits to Tourist Attractions 1994* (London: BTA/ETBRS, 1995); *Sightseeing in the UK 1995* (London: BTA/ETBRS, 1996); *Sightseeing in the UK 1996* (London: BTA/ETBRS, 1997); *Sightseeing in the UK 1997* (London: BTA/ETBRS, 1998); *Sightseeing in the UK 1998* (London: English Tourism Council/Northern Ireland Tourist Board/Scottish Tourist Board/Wales Tourist Board, 1999); *Sightseeing in the UK 1999* (London: English Tourism Council Research and Intelligence, 2000); Moffat Centre for Travel and Tourism Business Development, *Sightseeing in the UK 2000* (London: ETC/NITB/STB/WTB, 2001); VisitBritain, *Visitor Attraction Trends*

*England 2003* (London: VisitBritain, 2004); *Visitor Attraction Trends England 2005* (London: VisitBritain, 2006); *Visitor Attraction Trends in England 2007* (London: VisitBritain, 2008).

19  See Nick Laister, this volume.

20  Minutes of the Southport Area Committee (of Sefton Council) (4 October 2006): http://modgov.sefton.gov.uk/moderngov/Data/Southport%20Area%20Committee/20061101/Agenda/Item%2002.pdf accessed 16 November 2013.

21  Jo Kelly, 'Bid to Settle Row over Damage at Pleasureland out of Court', *Southport Visitor* (22 December 2008): http://www.southportvisiter.co.uk/southport-news/southport-southport-news/2008/12/22/bid-to-settle-row-over-damage-at-pleasureland-out-of-court-101022-22514825/ accessed 27 November 2011.

22  Using postings on web forums raises certain ethical issues. The analysis follows the guidelines issued by the British Psychological Society, *Ethics Guidelines for Internet-Mediated Research* (Leicester: British Psychological Society, 2013): http://www.bps.org.uk accessed on 21 November 2013. The SouthportGB forums are open and public (although registration is needed to search or post to them). No vulnerable groups were identified among the posters and there was little prospect of harm coming to any of them through citation here of their posts. In accordance with the BPS guidelines no online identities of forum uses are given nor the url from which quotes were taken. Furthermore, all quotes in the following section have been slightly altered or paraphrased so that they cannot be identified by search engines.

23  See, for example, Allan Brodie and Gary Winter, *England's Seaside Resorts* (London: English Heritage, 2007); John K. Walton and Jason Wood, 'Reputation and Regeneration: History and the Heritage of the Recent Past in the Re-making of Blackpool', in Lisanne Gibson and John Pendlebury (eds), *Valuing Historic Environments* (Farnham: Ashgate, 2009), pp. 115–37.

24  Laurajane Smith, *Uses of Heritage* (London: Routledge, 2006), pp. 29–33.

25  Laurajane Smith and Emma Waterton, '"The Envy of the World?": Intangible Heritage in England', in Laurajane Smith and Natsuko Akagawa (eds), *Intangible Heritage* (London: Routledge, 2009), p. 291.

26  See Allan Brodie and Roger Bowdler, this volume.

27  Brian Graham, G. J. Ashworth and J. E. Tunbridge, *A Geography of Heritage: Power, Culture and Economy* (London: Arnold, 2000), pp. 204–207.

28  Esteban Ruiz Ballesteros and Macarena Hernández Ramírez, 'Identity and Community – Reflections on the Development of Mining Heritage Tourism in Southern Spain', *Tourism Management* 28 (2007), p. 680.

29  Harrison, 'What is Heritage?' pp. 8–9; Harrison, *Heritage: Critical Approaches*, p. 15.

30  Sheila Watson, 'History Museums, Community Identities and Sense of Place', in Simon J. Knell, Suzanne MacLeod and Sheila Watson (eds), *Museum Revolutions: How Museums Change and Are Changing* (London: Routledge, 2007), pp. 163–68.

31  Smith and Waterton, 'The Envy of the World', pp. 294–5.

32  Dean MacCannell, *The Tourist: A New Theory of the Leisure Class,* rev. edn (New York: Schocken Books, 1989), p. 41.

33  Northwest Development Agency, *A New Vision for Northwest Coastal Resorts* (Warrington: Northwest Development Agency, 2003): http://www.coastalcommunities.co.uk/library/published_research/A_new_vision_for_Northwest_coastal_resorts.pdf accessed 25 November 2013.

34  Northwest Development Agency, *A New Vision*, p. 4.

35  Ibid., p. 66.

36  Ibid., p. 29.

37  Ibid., p. 30.

38  Ibid., pp. 67–8.

39  Locum Destination Consulting, *Southport Classic Resort Brand Study* (Haywards Heath: Locum Destination Consulting, 2004): http://modgov.sefton.gov.uk/moderngov/

Data/Cabinet%20Member%20-%20Regeneration/20040616/Agenda/Appendix%20 04A.pdf accessed 21 November 2013.

40  Ibid., p. 23.

41  The Southport Partnership, *Aspiring to be England's Classic Resort* (Southport: The Southport Partnership, n.d.).

42  David Bartlett, 'Plans to Transform Southport Pleasureland Abandoned", *Liverpool Echo* (9 February 2012): http://www.liverpoolecho.co.uk/news/liverpool-news/plans-transform-southport-pleasureland-abandoned-3352285 accessed 25 November 2013.

43  BBC News, 'Development Plan for Fairground': http://news.bbc.co.uk/1/hi/england/ merseyside/6529461.stm accessed 25 November 2013.

44  'Urban Splash Scoops Bid for £80 Million 'Classic Resort", *Liverpool Daily Post* (21 March 2008): http://www.liverpooldailypost.co.uk/news/liverpool-news/urban-splash-scoops-bid-80m-5515265 accessed 25 November 2013.

45  David Bartlett, '£80m Plan for Southport's Pleasureland Abandoned', *Liverpool Daily Post* (20 February 2012): http://www.liverpooldailypost.co.uk/news/liverpool-news/80m-plan-southports-pleasureland-abandoned-5440172 accessed 25 November 2013.

46  Katie Oakes, 'New Southport Pleasureland Here to Stay After £80m Transformation Plans Fall Through', *Southport Visitor* (9 February 2012): http://www.southportvisiter. co.uk/southport-news/southport-southport-news/2012/02/09/new-southport-pleasure-land-here-to-stay-after-80m-transformation-plans-fall-through-101022-30292946/2/ accessed 25 November 2013.

47  Joe Thomas, 'Southport Pleasureland's Owner Norman Wallis Talks about the Resort's Future', *Liverpool Echo* (21 March 2013): http://www.liverpoolecho.co.uk/news/ liverpool-news/southport-pleasurelands-owner-norman-wallis-3010391 accessed 27 November 2013.

48  Sefton Council, *A Draft Plan for Sefton, Preferred Option July 2013* (Southport: Sefton Council, 2013): http://www.sefton.gov.uk/pdf/Chapter%208%20Sustainable%20Growth% 20and%20Regeneration.pdf accessed 27 November 2013.

49  There is already a destination park – Blackpool Pleasure Beach – near to Southport and, as Table 9.1 shows, its visitor numbers show a long-term decline. Moreover, one of the north-west's theme parks – Camelot, a site located close to the M6 motorway and easily accessible from Liverpool and Manchester – closed at the end of 2012 due to falling visitor numbers.

# 10 Delivering the dream

## Saving Britain's amusement park heritage and the reawakening of Margate's Dreamland

*Nick Laister*

The escalation in property values and house prices that started in the UK in late 1995, and ended so abruptly in 2008, was one of the longest periods of continuous growth that the country has experienced in living memory. This economic cycle created significant problems in so many areas. One that has quite understandably been overlooked is the impact that the property 'boom' had on Britain's seaside amusement parks. A sizeable proportion of these parks closed between 1995 and 2007, when their owners – many of whom were second or third generations of the founding families – came to realise that the value of the land, often in suitable locations for housing and retail development, offered greater economic returns than the relatively limited profits being generated by their businesses. These closures accelerated almost in perfect unison with the rise in land values.

That is not to say that the closure of seaside amusement parks was caused solely by what was happening in the UK economy; the reality was that the growth in the value of property accelerated a process that had already started. Britain's seaside resorts had been declining for many years due to changes in the way people took their holidays; an accelerating trend well documented by the English Tourism Council (ETC).[1] Contributing factors included the rise of package holidays, cheap flights and competing domestic destinations from the 1970s and, importantly in this context, the growth and further development of inland resort destinations such as Alton Towers, Drayton Manor and Chessington World of Adventures in the 1980s. The result was that British seaside resorts found themselves with too much infrastructure, designed decades earlier to support visitor levels that were no longer being achieved. Amusement parks were part of that infrastructure.

A separate issue generally affecting British amusement parks during the 1970s and 1980s was the impact of the tragedy at Battersea in 1972. On 30 May of that year, a packed train, full of mostly youngsters on their Whitsun holidays, was climbing the first lift hill of the park's Big Dipper rollercoaster (actually a traditional Scenic Railway, similar to that at Margate's Dreamland), when the steel cable that pulled the train snapped and the train hurtled backwards down the incline (the secondary mechanisms that should have prevented this also having failed). Three children were killed and 16 others were injured.[2] Following this, a number of wooden rollercoasters in other amusement parks were removed.[3]

As a planning and development consultant working within the visitor attractions industry and involved in a number of seaside regeneration projects and advising a large proportion of the UK's amusement park operators on planning matters, I began to take a close interest in the closure of seaside amusement parks from the 1990s onwards. I was concerned about the impact these closures had on local tourism economies and especially the loss of the historic rides, none of which was recognised either statutorily or academically. I set about researching the histories of these parks and photographing the rides, buildings and other attractions.[4]

In 2001, midway through the property boom, one of the most significant and well known of Britain's historic amusement parks, Dreamland in Margate, came under threat. This was worrying as the park was home to the UK's oldest surviving rollercoaster – the 1920 wooden Scenic Railway – and, almost certainly, the oldest amusement park in the UK surviving on its original site.[5]

Although it only became a modern, 'American-style' amusement park from 1920, Dreamland's history as an amusement park can be traced back to 1867, when catering contractor Spiers and Pond leased a disused railway building from the London, Chatham and Dover Railway.[6] The railway building, renamed 'Hall-by-the-Sea', was initially used as a restaurant by day and a dance hall by night. It could not compete, however, with the well-established Assembly Rooms in the town centre, and in 1870, after only operating the Hall for three years, Spiers and Pond pulled out of its lease and the building was sold to Alderman Thomas Dalby Reeve, the then Mayor of Margate. In 1872, Reeve's son Arthur married Harriett Sanger, daughter of famous Victorian showman and self-ennobled 'Lord' George Sanger.[7] Sanger and Reeve jointly agreed to develop the Hall-by-the-Sea, which included a large area of land to the rear.[8]

The building, although refurbished, remained in essentially the same use as it had been under Spiers and Pond, with Reeve taking responsibility for the management. But it was Sanger, who laid out the land to the rear of the building as pleasure gardens, who can be credited with commencing the site's development as an amusement park. The pleasure gardens featured a menagerie of animals, lakes, statues, a mock ruined abbey and various amusements, including a waxworks, a steam-powered roundabout and an array of swings, archery ranges and a coconut shy all run by Sanger's less famous brother William.[9] In May 1874, Sanger advertised that the Hall-by-the-Sea, Italian and Zoological Gardens at Margate would open for the season, from the Whit Monday:[10]

> The Management beg respectfully to call attention to Italian and Zoological Gardens. Neither money, trouble, or taste has been spared in making these the most enjoyable Promenade by the Sea-side. The aid of Flora has been successfully invoked, and principal treasures in Trees, Shrubs, Plants, Flowers, and Statuary are represented. The Zoological Collection will comprise specimens of the most varied kind, and will constantly receive important additions of all kinds of birds and beasts.[11]

Sanger acquired the freehold to the Hall in 1875 following Thomas Dalby Reeve's death and further developed the pleasure gardens until his own death in 1911.

Hall-by-the-Sea struggled on, under the ownership of Arthur and Harriet Reeve, but it failed to capture the public's imagination and in 1919 they took the decision to sell the site.

The new owner was John Henry Iles in a joint venture with showman C. C. Bartram. Iles, who ran an advertising agency, was also passionate about brass bands, and during world tours with his band had witnessed the popularity of American amusement parks. It is reported that in 1906, during a visit to New York's Coney Island, he obtained the European rights to the Scenic Railway rollercoaster.[12] A year later, he built the UK's first Scenic Railway at Blackpool Pleasure Beach. At over £15,000, it was reputedly more costly than any at that time in America.[13] Iles also built other Scenic Railways at exhibitions in London (White City in 1908 and Earl's Court in 1909) and set up a number of amusement parks in Paris, Brussels, Petrograd, Cairo and Berlin. But Dreamland, the name Iles gave to his new Margate investment, was his first complete amusement park in the UK, opening in 1920 with its centre-piece Scenic Railway.[14]

The appearance of the park at this time suggests that it was partly modelled on Blackpool Pleasure Beach. Visitors in the 1920s would have experienced modern and thrilling variations on the roundabout, such as the Caterpillar (Figure 10.1) and Tumble Bug, traditional fairground rides such as the Helter Skelter and Gallopers, along with more elaborate permanent attractions, such as the House of Nonsense, Miniature Railway, Zoo (Figure 10.2) and River Caves, in which passengers were taken on a journey around the world in circular tubs that drifted sedately along an indoor waterway. Other rollercoasters operating during the 1920s were the Racing Coaster (a Figure Eight side friction coaster) and the Great Whirlwind, a twin-track racing coaster (Figure 10.3). Numerous sideshows were erected along the perimeter of the site, with brightly painted facades disguising primitive, and sometimes makeshift, sheds behind.

Following Iles' bankruptcy in 1938, the result of an unsuccessful flirtation with the film industry, Dreamland was taken over by his son Eric, but soon closed following the outbreak of the Second World War when the park went into voluntary liquidation. It reopened in June 1946, following an injection of cash from Billy Butlin, who by that time had established several of his famous holiday camps. Butlin himself was Chairman of Dreamland from 1946 to 1950 (Figures 10.4 and 10.5). In the 1980s, the park was taken over by the Bembom Brothers, and briefly renamed Bembom Brothers Theme Park. In 1995, the park was purchased by showman Jimmy Godden who also owned the Rotunda Amusement Park at Folkestone. Godden, operating the park as Dreamland Leisure Ltd, refurbished the entire park, including the Scenic Railway, and removed many of its existing permanent rides, including the Big Wheel, Looping Star rollercoaster and traditional Water Chute.

## The Save Dreamland Campaign

My direct involvement in Dreamland dates from May 2001, when I wrote to the government's Department for Culture, Media and Sport and requested that the

*Figure 10.1* The Caterpillar with the Scenic Railway beyond, 1920s

Source: © Joyland Books Archive

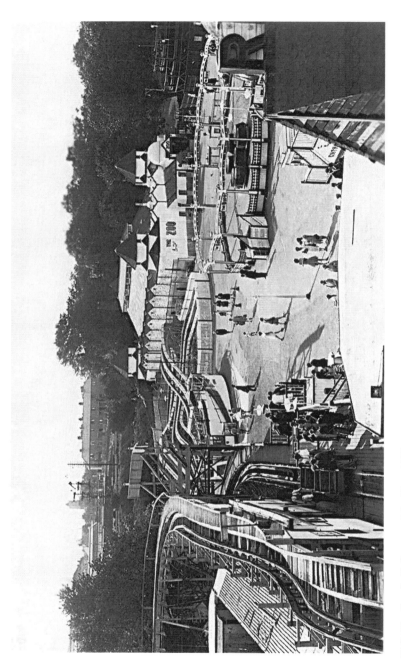

*Figure 10.2* The Scenic Railway and the Zoo, 1920s

Source: © Joyland Books Archive

*Figure 10.3* Various rides and arcades with the Racing Coaster beyond, 1920s

Source: © Joyland Books Archive

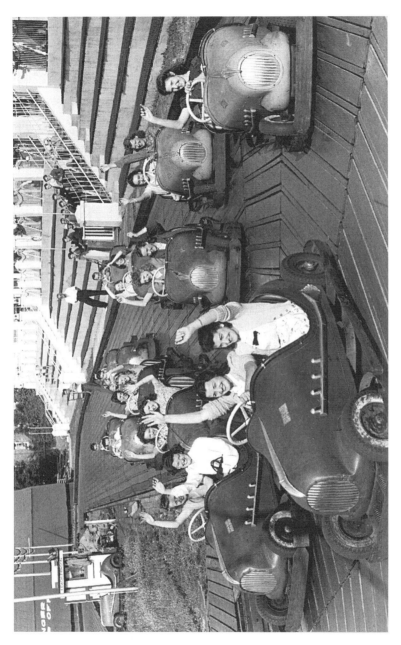

*Figure 10.4* A publicity shot of the Brooklands Racers, late 1940s

Source: © Margate Museum

*Figure 10.5* Various rides and stalls with an arcade in the shape of the liner Queen Mary beyond, early 1960s

Source: Phil Gould © Joyland Books Archive

Scenic Railway rollercoaster be spot listed.[15] My report provided background context to the Scenic Railway, showing that it was a remarkable and rare survivor of a ride that was once common in the UK. It noted that the Scenic Railway could be under threat due to the fact that a third of Dreamland had been cleared and I had heard rumours that the owner, Jimmy Godden, may have been looking to close the park. I stated that the ride should be protected, as there was a strong likelihood it would otherwise be lost to redevelopment. In 2002, the ride was listed Grade II, the first amusement park ride to gain this designation.[16] This resulted in a brief flurry of media interest but linked my name with Dreamland, which was to be instrumental in my increased involvement with the future of the amusement park a year later.

In early January 2003, the local press confirmed that the owner of Dreamland was indeed considering closing the park and redeveloping it.[17] As a resident of Oxfordshire, I was not immediately aware of this, but soon started to receive a number of emails from people who, remembering my involvement in the listing of the Scenic Railway, asked if I could help, as they were worried about the effects that the closure of Dreamland would have on Margate's tourism economy. Within days I decided to set up the Save Dreamland Campaign, initially as a small website on which to gather names and to allow me to make representations to Thanet District Council to ensure that the park was appropriately protected in the Local Plan. My initial approach was to lobby the Council, partly on the basis that planning policy guidance at the time required owners of listed buildings to provide evidence that a site containing a listed building is no longer commercially viable, including a requirement to market the unrestricted freehold on the open market. I also pressed the Council to ensure that the site was properly protected as an amusement park in the Local Plan, to ensure that the hope value of the land did not price out any future park operators. At this stage of the Campaign I regularly made reference to the Thanet Local Plan Deposit Draft 2001, which protected the Dreamland site as an amusement park and stated that proposals that would lead to a reduction in the attractiveness or tourism potential of Dreamland Amusement Park would be resisted. It allowed in exceptional circumstances the development of a limited part of the site, provided the future viability of the amusement park could be assured.[18]

A number of organisations joined the Campaign in the early weeks, including SAVE Britain's Heritage, the Margate Historical Society, the Margate Civic Society, the Margate Hotel and Guest House Association, the Fairground Association of Great Britain, American Coaster Enthusiasts, the European Coaster Club and the Roller Coaster Club of Great Britain.

In an interview published in February 2003, Jimmy Godden stated that he was closing Dreamland to enable him to 'take life easier'.[19] The article cited the 'commercial value of the land' as one of the reasons behind the decision to close both Dreamland and the Rotunda in Folkestone at the end of the 2003 season. Godden admitted that if he were to sell either Dreamland or the Rotunda as a going concern, 'the price I would sell at would have no bearing as to their true commercial development land value'. He also revealed that his plans were a 'purely commercial decision', and added that 'it may well be the start of things to come when you are looking at a lot of coastal amusement parks and their development values'.

The Dreamland site found itself vulnerable to attack from vandals and arsonists almost immediately. In April 2003, a large fire destroyed Mr G's Amusement Arcade on the seafront, as well as damaging the neighbouring Funland arcade. These arcades backed onto the Dreamland site. Mr G's, as the name implies, was owned by Jimmy Godden. Later in 2003, a fire broke out on the Waltzer ride, completely destroying it.

After Dreamland Leisure stopped operating the park at the end of 2002, it struggled on with a line of travelling fair operators taking out leases, but with its rides slowly being sold off each year, the park diminished in size. Nevertheless, it was still seen as a going concern to some. The Save Dreamland Campaign was contacted by two theme park operators interested in acquiring the site and operating it as a modern amusement park. The first of these was Grévin and Cie, a Paris-based theme park operator with a number of sites around Europe (including the famous Parc Asterix). They visited in April 2003 and made an offer to acquire the site. But Godden refused this offer because it was below his estimation of its residential and retail development value. The second operator was the owner of Southend's successful Adventure Island amusement park, who contacted the Campaign in November 2003.[20] Adventure Island made a similar offer and again this was refused. This reinforced the Campaign's belief that it was necessary for the Local Plan to protect Dreamland; otherwise it was unlikely that any tourist attraction operator would be able to compete with the values that were being sought for residential or retail development.

The operator of Dreamland in the 2003 season, David Wallis, also told the local press that he believed Dreamland to be viable[21] and confirmed that he would be interested in operating the park again for the 2004 season.[22] And in a separate interview, David Collard, who operated the Scenic Railway independently throughout the 2003 season, stated: 'It worked out very well and easily paid for itself, with quite a bit left over'.[23] At the time of this news, I made a statement on the Save Dreamland Campaign website, which summed up the position as I saw it at the time:

> This is yet more good news for Margate. We already knew that more than one established theme park operator wants to buy the site and invest in its rides and attractions. We now have further confirmation that this important tourist attraction is a viable business. The very person who operated it this year has now confirmed that he would do it again next year. He would not be doing that if the park doesn't make money.
>
> And further confirmation from Dave Collard that the Scenic Railway, operated as a separate concession, can more than pay its way is very welcome news indeed . . . We now know the park is viable; that is no longer the issue. *Park World* correctly identifies that land values are the only issue remaining. We now call on the Council – with all the evidence now available to it – to stand firm in its rejection of retail on the site.[24]

David Wallis would operate the park again in 2004. Interviewed after his second season, he again provided evidence of the park's viability, saying that he would like to return for another season if owner Jimmy Godden agreed, adding that: 'We do feel that there is a need for this type of entertainment in Margate'.[25]

Dreamland reopened for the 2005 season but David Wallis was not able to agree terms with the owner, so the park was leased to another showman, Harry Ayers. Many considered this to be one of Dreamland's worst seasons for many years, with the park's appearance looking poor, and its opening dates erratic. Its final year of operation was in 2006, when the park was leased by another show-man, George Webb. The park finally opened at the end of May but struggled to work due to there being only a very small selection of rides – with the exception of the Scenic Railway, all of the permanent rides had now been sold – and due to the fact that the main entrance was closed all season so visitors had to enter by walking around the back of the amusement arcades. This was a very unsatisfactory year for the park, and would also be the final year that the Scenic Railway operated.

With the backdrop of dwindling rides and seemingly little interest in a long-term future for the site as an amusement park, the Campaign's approach to saving Dreamland was to use the planning system, and in particular the emerging Local Plan. In May 2003, the Campaign submitted formal representations to the latest version of the Thanet Local Plan, objecting to the changes to Policy T11, which protected the site as an amusement park.[26] The representations stated that the Policy as previously drafted ensured that the Dreamland site remained primarily as an amusement park for the benefit of Margate's tourism economy, and that the results of research undertaken by the Campaign suggested that the amusement park was viable and that a redevelopment that does not include an amusement park element is completely unnecessary (as well as potentially harmful to the economy of the town). The Campaign also presented its own vision for the site – a vision informed by discussions with major theme park operators.

The strength of feeling in the town about the future of Dreamland was illustrated by the fact that changes to Policy T11 received by far the most objections of any proposed policy in the Plan, demonstrating the lack of public support for the Council's amendments. However, despite the strength of the objections, the Council made no immediate change to the amended Dreamland policy, stating that it was awaiting the results of a study by consultants.

The Thanet Local Plan Inquiry was held in spring/summer 2004 and was to prove a turning point in the Save Dreamland Campaign. I submitted a proof of evidence on behalf of the Campaign and gave evidence at the Inquiry. Susan Marsh (who would later become the Secretary of The Dreamland Trust) advocated for the Campaign. In my proof of evidence,[27] I made the following key points:

- The Dreamland site was of critical importance to Margate as a seaside resort, and must be retained and enhanced. It was the only tourist attraction in the Isle of Thanet that drew more than 100,000 visitors (even in its then artificially run-down state it drew almost 700,000 visitors per year in 2002, placing it in the top ten amusement parks in the country). I presented evidence that gave some initial indication of the effects of Margate's first Easter without Dreamland (9 to 12 April 2004), with disappointment expressed by visitors and tourism businesses.

- I set out the positive impact that an amusement park can have on a seaside resort. I used the example of Adventure Island at Southend, which opened in 1998. At Southend, surrounding tourism businesses all responded to the investment in a successful amusement park by investing in their own facilities. I explained that the changes in the overall appearance of Southend's seafront area since it opened had been marked and that the fortunes of the town had been completely turned around by the continued investment in the amusement park over this period.
- I also made an important distinction between a tourism land use, which attracts people to a town, and other uses such as retail and leisure, which primarily serve the local population (albeit that they also provide ancillary facilities for tourists).
- I presented evidence on the heritage of the Dreamland site, which was unique in this country and should therefore be protected. The Scenic Railway rollercoaster was the UK's oldest operating rollercoaster and was considered to be of international importance.
- I also explained that there was interest from established amusement park operators in acquiring and investing in the park. I presented a significant amount of evidence on this. I had no doubt that Dreamland could not only survive but also prosper under one of these interested operators. I stated that there was absolutely no reason why Margate should lose its biggest tourist attraction, as long as planning policies continued to protect it for this use.

The independent Inspector's Report on the Thanet Local Plan Inquiry was published in November 2005. The Inspector, Harold Stephens, accepted every point made by the Campaign and rejected virtually every point made by the Council and the site's owner. He found the evidence of the Campaign to be compelling and asked the Council to change the Local Plan to ensure the protection of the amusement park and the listed Scenic Railway rollercoaster. The Inspector described the Scenic Railway as an 'extraordinary building' and was satisfied that it would be viable, even as a stand-alone attraction. Importantly, however, he stated that the setting of the Scenic Railway must also be conserved, thereby preventing the site from being used for anything other than an amusement park.[28]

However, the Council's Planning Policy Manager recommended that the Council partially reject the Inspector's conclusions, allowing Dreamland to be redeveloped if an amusement park was not viable. The Council voted this through and went out to consultation on that basis. A report on the consultation was published in May 2006, covering all aspects of the Plan.[29] It stated that the consultation prompted 452 responses, of which 442 were concerning Dreamland, and the vast majority of those were objections from people asking for the Inspector's recommendations to be accepted in full. This demonstrated the extent to which, after three years of campaigning, the people of Margate were still engaged in this process.

At a lively meeting of the full Council on 11 May 2006, the policy as proposed by officers was approved but with modifications to ensure that proper consideration was given to Dreamland's viability should an attempt be made by the owners to demonstrate that the park was unviable. The Campaign spoke to the press and confirmed that they were satisfied that the Council had approved an appropriate policy for Dreamland and confirmed its willingness to work with the Council. The first seeds of bringing back Dreamland had been sewn.

However, between the Inquiry and the publication of the Inspector's Report, the Dreamland site was sold for an undisclosed sum (reported in March 2005 by the BBC to be '£20m') to the Margate Town Centre Regeneration Company (MTCRC). Its chairman, Toby Hunter was quoted as saying:

> It is going to be a fantastic mix of leisure, restaurants, retail and accommodation. . . . We want the site to promote Margate for the future rather than it be retrospective. We just have to see whether the Scenic Railway is too much of a retrospective.[30]

The Campaign learned of MTCRC's approach to the site and the Campaign itself in a press interview conducted with Hunter almost a year later:

> Save Dreamland Campaign has not invested a single penny into Margate – but it makes a lot of noise. I would welcome any effort this campaign can make towards investing in Margate. Ultimately, regeneration can only happen through investment. My business partner and I have, so far, invested £9.5 million in Margate and I intend to invest more.[31]

Hunter may have been unaware that in March 2005 the Campaign had tried to set an agenda for the future of Dreamland by launching its own vision statement, put together in association with the French theme park designer Jean-Marc Toussaint and Southend's Adventure Island Theme Park. The concept, based on how the park might look if taken over by an established operator such as Adventure Island, showed the sort of rides and attractions that might be found in the park in the near future if the Campaign was successful.[32] At the time, despite a lot of support from the public, this did not gain much traction. Clearly, following the successful Inquiry and the realisation after the end of the 2006 season that the park was not likely to open again, a different approach was going to be needed.

It was at this point in the story, in early 2007, that I came to the conclusion that it was necessary to do more than just sit on the sidelines and that it may be required to get directly involved in bringing Dreamland back. There was a new owner and we had reached agreement with the Council on a policy for Dreamland in the Local Plan, but the park was to stay closed.

## The world's first heritage amusement park

A meeting on 15 January 2007, at which Sarah Vickery of the Save Dreamland Campaign and I met with representatives of Margate Renewal Partnership (MRP),[33] Thanet District Council and Locum Consulting, proved to be the big turning point for the Campaign and the future of Dreamland. Sarah and I put forward a proposal to work jointly with the Council (and hopefully the new owners) to create the world's first heritage amusement park, using the newly adopted Local Plan policy for Dreamland as a springboard. Our proposal was to create an amusement park of historic rides. I had been inspired by the spate of seaside amusement park closures in the period up to 2007. As mentioned earlier, a number of park operators (some my own clients) were in the process of closing their parks and selling the land to property developers. Some of these parks included historic rides, many being the last surviving examples of their type: for example, Frontierland, Morecambe (closed 1999); Spanish City, Whitley Bay (closed 2000); Rotunda Amusement Park, Folkestone (closed 2005); and Pleasureland, Southport (closed 2006).[34] Clearance of Southport's Pleasureland site was imminent and I was also aware that Rhyl's Ocean Beach Amusement Park, the biggest seaside amusement park in Wales, was to close by the end of the year, with the loss of a rare Water Chute ride. In conclusion, I pointed out that, outside of Blackpool and Great Yarmouth, virtually all of the UK's amusement park heritage would be lost in the space of the next few months, such was the pace of property development at this time.

I saw an opportunity to reopen Dreamland and to create a unique heritage attraction set around the country's oldest rollercoaster in the oldest surviving seaside amusement park. If we could rescue the rides that were about to be destroyed at amusement parks around the UK coast, we could bring them together to create a really distinctive attraction that could contribute significantly to Margate's regeneration. I proposed that we approach MTCRC to see if we could find a way of working with them on the project, to enable them to secure some redevelopment on the site and at the same time enable Dreamland to reopen in accordance with the Local Plan. I saw this as a partnership between the Save Dreamland Campaign, the Council and MTCRC. I also believed it was something that would meet the aspirations of the public, businesses and other stakeholders that had engaged so strongly in the debate on the future of Dreamland over the past four years.

Our proposal was well received by those at the meeting. I agreed to draw up a report on the UK's remaining amusement park history, identifying those rides that were immediately under threat, and put together my thoughts on how we could make this happen. I also agreed to start thinking about a vehicle with which we could deliver this, as it was likely that the Campaign would need to become directly involved. It was at this point that Sarah and I, along with Susan Marsh (who had played such a central role at the Local Plan Inquiry) decided to set up The Dreamland Trust as a vehicle to raise funds and potentially get engaged in the operation of Dreamland and/or the Scenic Railway. The Trust was formed on 2 February 2007 by the three of us, and eventually became a not-for-profit limited company.

After further development of the concept of a heritage amusement park, I produced a confidential report on historic rides that were potentially available for the project, along with some very initial work on viability.[35] We did not make a particularly good start in acquiring rides. In early April 2007, I worked tirelessly to secure the 1922 Runaway Coaster at Folkstone's former Rotunda Amusement Park, it being the only side friction Figure Eight coaster left in the UK and one of only two in the world (similar to the Racing Coaster that once operated at

Dreamland). Yet, despite agreeing terms in principle with the owner, this important part of amusement park heritage was demolished.

Meanwhile, the Save Dreamland Campaign launched its updated vision, this time for an amusement park of historic rides. I made the following statement on our website on 30 March 2007:

> Our March 2005 Concept Plan remains our favoured option for the site. However, without the strong support we wanted for the protection of Dreamland in the Local Plan, we have had to think of a 'Plan B', which allows for some of Dreamland to be redeveloped whilst ensuring the continuation of a major visitor attraction on the site. The Heritage Amusement Park is our Plan B – but what an exciting Plan B it is.
>
> The Save Dreamland Campaign has held several meetings over the past few months with Thanet District Council and the Margate Renewal Partnership . . . to develop the concept of a Heritage Amusement Park at Dreamland, based around the listed Scenic Railway. The Heritage Amusement Park, which would be a world's first, would include some of the remaining examples of Britain's amusement park heritage, in a high-quality park-like environment around the Scenic Railway. The listed Cinema building would also be brought back into use with rides, shows, bars, restaurants and an amusement park/seaside heritage museum.
>
> The Campaign has carried out some initial feasibility work on the viability of the project and commissioned theme park designer Jean-Marc Toussaint to produce a new Concept Plan for the site showing how the park could look. The Plan showed a potential selection of vintage amusement park rides of the type that could operate at the park.[36]

The Campaign's position was that some of the park could be redeveloped if it ensured its long-term survival. Funding from the developer, along with other grant funding, could help to deliver this attraction. The park could be owned and potentially operated by a trust, and would be self-sufficient. The Campaign was confident that this was a realistic proposition that should attract hundreds of thousands of visitors to Margate, and offered to work with MTCRC in bringing a heritage amusement park to Dreamland.

There was extensive media coverage of the proposals in the weeks following the launch and the feedback received was overwhelmingly positive. Thanet North MP Roger Gale came out in favour, stating that he owed the Save Dreamland Campaign a 'debt of gratitude', adding that without the Campaign's enthusiasm 'it is unlikely that the present concept of a Heritage Park would have been included in one of the plans'.[37]

Things were moving quickly. I was now in talks with the owners of several closed or soon-to-be closed amusement parks or rides around the UK about taking a small selection of the rides that best represented Britain's amusement park heritage. This soon resulted in the acquisition of various rides from Southport's Pleasureland, Rhyl's Ocean Beach Amusement Park, and Blackpool Pleasure Beach.

Work began at Pleasureland where a number of vintage rides like the River Caves and Caterpillar, both dating back to the 1920s, were due to be demolished in June 2007 to make way for a temporary fairground operator and ultimately a completely new redevelopment. I had been in discussions with Pleasureland's new owners, Sefton Council, and the site's former owners, Blackpool Pleasure Beach, since February 2007 about the rides being donated to the Dreamland project. This was finally agreed. The rides acquired included the 1922 Caterpillar (identical to the one that operated at Dreamland) and the 1960 Wild Mouse wooden rollercoaster that had originally operated at Morecambe's Frontierland. We also acquired a Flying Scooters ride, the Pleasureland cable car system and Ghost Train systems, the first meteorite ride to operate in a UK amusement park and several machines from the Fun House. Dismantling and transportation costs were met by MTCRC.

In October 2007 came news that the UK's last surviving circular Water Chute at Rhyl's defunct Ocean Beach Amusement Park had been demolished. Fortunately it was possible to salvage the key remaining mechanical parts of the ride – the motor, gears, pumps, etc –along with the boats. This meant that a complete rebuild of the ride became a possibility. We made an appeal for funds to pay for their transportation and very quickly secured the required £1,500. In 2008, following several months of negotiation, the UK's only surviving full-size Whip was acquired. The ride, which had operated at Blackpool Pleasure Beach since at least 1921, was manufactured by W. F. Mangels of Coney Island and was identical to the Whip that once operated at Dreamland. Again, the dismantling and transportation costs were met by MTCRC. The following year we also acquired the Junior Whip from the Pleasure Beach (Figure 10.6).

As the collection of historic rides steadily grew, Susan Marsh and I met with Heritage Lottery Fund (HLF) in 2007. This was an initial discussion on the prospects of The Dreamland Trust making a successful application for funding to assist with the heritage amusement park proposal. Also in attendance at the meeting were Doug Brown and Nick Dermott, representing the Council and MRP's heritage consultant Jason Wood who had orchestrated the meeting.

Public support at this time had still not waned. A Council consultation on a new Planning Brief for Dreamland (adopted in February 2008) showed that there was still strong support for keeping the Scenic Railway (over 92 per cent of respondents wanted it retained – the strongest support for any part of the Brief), keeping more than half of the site as an amusement park (88 per cent), and for the Dreamland Cinema building to be kept for leisure uses (86 per cent). There was less support for the associated enabling development, including family homes and apartments, with 55 per cent disagreeing or strongly disagreeing with this idea. However, officers explained that this type of development was required to fund the provision and operation of a future high-quality amusement park and, in response to these concerns from local people, the brief was amended to explain this fully.[38]

A potential set back, however, was an arson attack on the Scenic Railway in April 2008 which destroyed about 25 per cent of the ride, along with part of the station, all of the workshops, and all of the trains (Figure 10.7). Nevertheless, Council Leader Sandy Ezekiel told the press that the Scenic Railway would be rebuilt.[39]

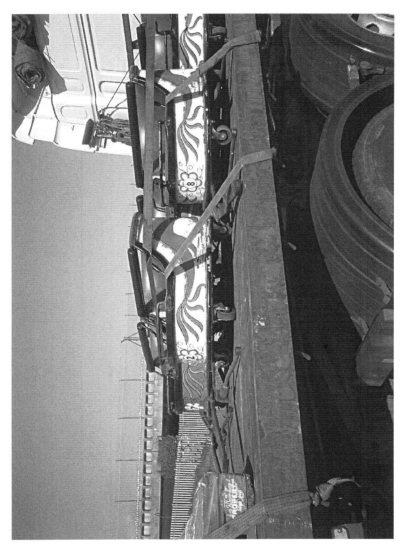

*Figure 10.6* The Junior Whip arriving at Dreamland from Blackpool Pleasure Beach, 2009

Source: Susan Marsh © The Dreamland Trust

*Figure 10.7* The fire damaged station of the Scenic Railway, 2008

Source: © Jason Wood

In June 2008, the Prince's Regeneration Trust (PRT) was jointly commissioned by MRP and MTCRC, in association with The Dreamland Trust, to undertake a feasibility study on the restoration of the Scenic Railway and the development of a heritage amusement park. This was the necessary first step to securing funding and hopefully bringing the project to reality. In a press release, PRT's Chief Executive Ros Kerslake said: 'Dreamland is a key site for Margate occupying a key location and having a central place in the public's perception of Margate as a Seaside town. There is a fantastic opportunity to celebrate the heritage of Margate while renewing its visitor economy'.[40] In their completed Scoping Report,[41] PRT indicated the areas of work needed to bring forward the regeneration of the 'heritage assets', proposed an outline programme and budget for the preparatory stages of the scheme, and set out a potential structure for project management. The report formed the basis of a funding application submitted by MRP to the government's Sea Change programme, which was designed to invigorate England's seaside towns through investment in culture and heritage.

Work then started in earnest. Clearance of the fire-damaged parts of the Scenic Railway commenced in September 2008 and in October MRP announced that its Sea Change bid for a grant of £30,000 had been successful, following which feasibility studies were undertaken and further funding applications prepared for HLF and Sea Change.

A key part the Dreamland Trust's activities at this time was meeting with key national, regional and local stakeholders in the hope that the project would also capture their imagination and support. One such organisation was the government's Urban Panel, who visited Margate in 2008. The Panel brought together the expertise of the Commission for Architecture and the Built Environment (CABE) and English Heritage with the aim of helping local authorities, development agencies and others to engage in major regeneration of historic towns and cities. The Panel's report, released the following year, identified Dreamland as the single most important project for the town:

> In choosing which of the many schemes to prioritise, the Panel had no doubts. Dreamland is blessed with assets of the highest quality and national importance, a nationally known (remembered) name, all the space that is needed, a dedicated Trust, with a collection of historic fairground rides and a vision which the Panel thought wholly apposite. Accordingly Trust, Partnership and authority were urged to turn all necessary attention to the re-emergence of Dreamland.

The Panel also believed that there was a very good case for funding for the project and recommended that it moved forward as a matter of urgency:

> National funding streams of relevance exist and, although the Panel has no lien on them, members thought there was very good case for some funds coming to Margate and being directed to Dreamland . . . So the Panel urged that the vision of Dreamland re-opening as the first and nationally unique

heritage fairground be promoted with urgency and drive and without unnecessary burdens. Similarly, the Panel saw no merit in delaying the scheme because the grander development package for the site (which suddenly looks very dated) cannot now be delivered. New housing may be a long time coming and yet this need not hamper successful delivery of new Dreamland. The remainder of the site can be brought forward, with minimal investment, as an events space. Panel members envisaged a near future in which Dreamland is functioning and attracting new visitors as well as entertaining and pleasing existing ones. That can be much enhanced by attracting to the new, large events space circuses, markets, small festivals and so on, with obvious concomitant benefits. The Panel was also confident that such a degree of activity would constitute the critical mass which would provide the context for a successful re-use of the cinema.

In concluding, the Panel even suggested that other projects should be abandoned or deferred to allow for Dreamland to move forward.[42]

After presenting the proposals for a heritage amusement park to the Board of MRP, and gaining the support of this key stakeholder group, the full scheme was finally launched on 15 March 2009 at the successful 'I Dream of Dreamland' event in Margate town centre. Over 400 people attended. In addition to the plans and drawings of what would be a 'world's first' attraction, there were various information boards, two separate film shows of Dreamland footage, models of various Dreamland rides, funfair food, usherettes and a Dreamland-themed disco.

Within weeks, The Trust submitted an application for £500,000 of development funding from HLF and MRP applied again to the Sea Change programme, this time for a much larger grant of £4 million. In July 2009, HLF gave the green light to the project when it awarded £384,500. This first-round pass meant that The Trust had up to two years to work up detailed proposals and apply for major funding.

A report in *The Observer*, one of a long line of national newspapers covering the project, succinctly summed up the project at that time:

> This idea, born of local enthusiasm and pride . . . seems to belong to Margate already – a privilege its new gallery [Turner Contemporary] will have to earn.
>
> I can think of few things lovelier than looking at a tempestuous Turner sky in the place where he painted it. But I also think that old wooden rollercoasters count as culture too . . . These things are part of who we are . . . To combine them, then, is very heaven.[43]

The most important landmark in the six-year campaign to save Dreamland came on 16 November 2009 when the government announced that it had awarded £3.7 million to the project. This was the largest grant in the 2009 Sea Change programme. CABE Chief Executive Richard Simmons, commenting on the seven coastal resorts that received a grant, said:

I especially like the plan to regenerate Dreamland in Margate, and showcase the country's oldest rollercoaster, a listed scenic railway. It is ambitious projects like this, creating new national attractions, that can rekindle the English love affair with our seaside.[44]

From that point, several additional funding sources began to open. The Council agreed to put up to £3 million and the HLF confirmed a further £3 million Stage 2 funding on 22 November 2011. In a statement, the head of HLF for the South East, Stuart McLeod, said:

This project has huge potential to entertain, intrigue and delight. As Margate's answer to an early Disneyland, the Heritage Lottery Fund is delighted to award this substantial investment to Dreamland and, once restored to its former glory, it will not only bring alive the historic rides of yesteryear but boost tourism too.[45]

Support of key stakeholders was also demonstrated by the upgrading of the listing status of both the Dreamland Cinema and Scenic Railway from Grade II to Grade II*. These amendments to their status were entirely appropriate recognition for these rare and vulnerable structures. It placed both into the top 6 per cent most important listed buildings in the country.[46]

## Delivering the dream

Despite all these funding pledges, the project could not be implemented as the Council and The Dreamland Trust were unable to reach agreement with MTCRC regarding the transfer of the amusement park site. As the funding that had been secured was dependent on the Council obtaining a long lease, the Council issued a Compulsory Purchase Order (CPO) on 3 June 2011. In the Council's press release I said:

We had hoped that the compulsory purchase of Dreamland would not be necessary, but any further delay could have jeopardised the project. Dreamland is an essential component of Margate's regeneration, being in such a prominent location on the seafront, and it will also provide training and employment opportunities for Thanet. We are therefore pleased that we can now move forward with our preparations for starting work on this world-first visitor attraction.[47]

A Public Inquiry into the CPO was to be the final big hurdle before work could start on the project. This got underway in January 2012 and ran, on and off, until March. In August, the Secretary of State approved the CPO, with the Inspector's report robustly backing the project. Unfortunately, this triggered a series of appeals, with the case being taken to the High Court and Court of Appeal. Whilst both courts backed the CPO, this created another twelve months' delay, with the site finally transferring to the Council in September 2013. This finally unlocked the funding and the project team's ability to complete its designs, make final ride acquisitions and appoint contractors to rebuild the Scenic Railway and carry out the site works. Additional rides acquired included the former Dreamland gallopers, a large adult jets ride built by Langwheels of Middlesex in the 1950s (identical to the one that operated in Dreamland for many years), and a number of juvenile rides.

In 2014, it became clear that the operating model that had formed the basis of the grant funding – the Dreamland Trust operating Dreamland on a not-for-profit basis – was not going to work. The Council's legal advisors said that Dreamland would have to be placed on the open market and the Trust would have to bid for the site. To meet the Council's procurement rules we would also have to demonstrate a level of financial stability that, despite our multi-million pound grants, we probably could not achieve. We took the very difficult decision not to bid to operate Dreamland, and to allow a commercial operator to take the lead. We felt that this was in the best interests of Dreamland, as the operator would have deeper pockets and expertise that we would simply not possess. We needed to ensure that Dreamland would be sustainable in the long-term, and history has shown that the private sector is best placed to operate amusement parks successfully.

A company called Sands Heritage was selected by the Council. Although the company was set up solely for the purpose of operating Dreamland (by the owner of Margate's Sands Hotel), the board had various industry veterans on it. The company worked incredibly hard to get the park open, whilst the Council and Trust worked on delivering the park to Sands ready for opening day.

Finally, after twelve years of leading the project to revive Dreamland, the park opened to the public. The big day was 19 July 2015. I gave a brief speech, before the red ribbon was cut by Heather Brennan, Henry Iles's great great granddaughter, George Weston Wright, former Scenic brakeman, and David Sanger and Caz Bartlett, descendents of Lord George Sanger.

The Trust had worked hard to make Dreamland more than just a collection of rides from different eras of amusement park development. We had appointed HemingwayDesign[48] to establish the branding and to give the park a distinctive look and feel, and work with the larger project team that involved architects, landscape architects, engineers, rides consultants and project managers. The Hemingway team had given it a really cool, slightly old-fashioned feel, which worked perfectly. There were lots of entertainments on offer, mostly free, and the inside of the entrance building perfectly captured the essence of a historic amusement park, complete with roller disco, vintage slot machines and a souvenir shop selling products made from the timber removed from the Scenic Railway.

The only cloud that hung over the launch was the sight of the Scenic Railway standing but not operating as the testing of the trains had not been completed. It would not operate until October. A number of other rides were non-operational as well, but temporary rides had been brought in to make up for that (in the spirit of seaside amusement parks of old). Despite these setbacks, the atmosphere was palpable and I received positive feedback from everybody I spoke to on the day.

The road to re-awakening Dreamland, and to preserving these important aspects of British amusement park history, has been a very long one. But hopefully it will result in this often overlooked part of the country's built and cultural heritage finally getting the recognition it deserves. And if Dreamland can play an important role in the ongoing transformation of Margate, then it will truly have been worthwhile.

## Notes

1 English Tourism Council, *Sea Changes* (London: English Tourist Board, 2001) is one of the first attempts by a government agency to understand the reasons behind the decline and to establish strategies for regeneration.
2 The Big Dipper was a rebuild of a much earlier ride that had operated at Southport's Pleasureland and at Sutton Coldfield's Crystal Palace funfair before being dismantled and reassembled at Battersea; see Ian Trowell, this volume.
3 For example, Barry Island Pleasure Park (Scenic Railway, 1973); Seaburn Fun Park (Big Dipper, 1973); Southend's Kursaal (Cyclone, 1974); Wonderland, Cleethorpes (Dipper, 1974); Ocean Beach Amusement Park, Rhyl (Big Dipper, 1975); Spanish City, Whitley Bay (Figure Eight, 1975); Manning's Amusement Park, Felixstowe (Giant Dipper, 1977); and Coney Beach Amusement Park, Porthcawl (Figure Eight, 1981). It is not possible to say whether in all these cases their removal was as a direct result of the Battersea accident: see Robert Preedy, *Roller Coasters: Their Amazing History* (Leeds: Robert Preedy, 1992).
4 As a chartered town planner, I use the planning definition of an 'amusement park' taken from the *General Permitted Development Order 2015*. This describes an amusement park as 'an enclosed area of open land, or any part of a seaside pier, which is principally used (other than by way of a temporary use) as a funfair or otherwise for the purposes of providing public entertainment by means of mechanical amusements and side-shows'.
5 The early history of Dreamland is well documented by Nick Evans, *Dreamland Revived: The Story of Margate's Famous Amusement Park*, rev. edn (Whitstable: Bygone Publishing, 2014).
6 In 1864, LCDR erected a seafront station building next to the existing station belonging to its rival, the South Eastern Railway, but the later station was never connected to the railway network as the anticipated parliamentary approval was never granted. LCDR later built a new station further to the west (Margate West, now Margate Station).
7 Preedy, *Roller Coasters:* p. 32.
8 This area was formerly a salt-marsh known as The Mere but after the Marine Terrace causeway was built the reclaimed land was used for planting potatoes: http://www.dreamland.org.uk/past.html accessed 27 June 2014.
9 Evans, *Dreamland Revived*, p. 7.
10 http://www.margatelocalhistory.co.uk/Pictures/Pictures-Dreamland.html accessed 27 June 2014.
11 Quoted by Antony Lee, 'The Reverend Wood Visits the Zoo', p. 2: http://www.margatelocalhistory.co.uk/Articles/wood.pdf accessed 27 June 2014. By 1876, the animal collection included lions, camels, leopards, a Bengal tiger, cranes and jackasses.
12 Preedy, *Roller Coasters,* p. 30.
13 Peter Bennett, *Blackpool Pleasure Beach: A Century of Fun* (Blackpool: Blackpool Pleasure Beach Ltd, 1996), pp. 24–5.
14 Iles later went on to operate a number of other amusement parks in the UK, including Great Yarmouth Pleasure Beach, Belle Vue in Manchester and Aberdeen Beach Amusement Park.

15  Nick Laister and David Page, 'Request for Spot Listing of the Scenic Railway Roller Coaster, Dreamland Amusement Park, Kent' (Wantage, 2001): available at www.save-dreamland.co.uk.

16  The Water Chute at Hull's East Park became a listed building a year later: http://www.britishlistedbuildings.co.uk/en-490436-water-chute-on-the-boating-lake-in-east-accessed 27 June 2014.

17  See, for example, *Isle of Thanet Gazette* (3 January 2003).

18  Thanet District Council, 'Isle of Thanet Local Plan, Deposit Draft' (Margate, 2001), esp. Policy T11, which protected the site as an amusement park.

19  *Park World* (February 2003).

20  Adventure Island, which opened in 1998 partly on the site of the former Peter Pan's Playground amusement park, had become Southend's main amusement park following the final closure of the Kursaal the previous decade.

21  *Isle of Thanet Gazette* (4 April 2003).

22  *Park World* (January 2004).

23  Ibid.

24  Save Dreamland Campaign website (8 January 2004): http://www.joylandbooks.com/scenicrailway/news0401.htm accessed 27 June 2014.

25  *Isle of Thanet Gazette* (3 September 2004).

26  Nick Laister, 'Revised Deposit Draft Thanet Local Plan: Representations of the Save Dreamland Campaign' (May 2003).

27  Nick Laister, 'Revised Deposit Draft Thanet Local Plan: Proof of Evidence of Nick Laister' (Margate: Save Dreamland Campaign, April 2004).

28  'Thanet Local Plan: Inspector's Report – Chapter 8 (Tourism)' (November 2005).

29  Thanet District Council, 'Report on Consultation on the Proposed Modifications to the Local Plan' (8 May 2006).

30  BBC News (11 March 2005): http://news.bbc.co.uk/1/hi/england/kent/4340993.stm accessed 27 June 2014.

31  *Isle of Thanet Gazette* (24 February 2006).

32  Nick Laister, 'Our Vision for Dreamland' (Margate: Save Dreamland Campaign, March 2005): http://www.joylandbooks.com/scenicrailway/vision.htm accessed 27 June 2014.

33  The Margate Renewal Partnership brought together public sector partners to secure the regeneration of Margate. The Partnership, chaired by the South East England Development Agency (SEEDA) Chief Executive, Pam Alexander, was established in 2006 and included representatives from Thanet District Council, Kent County Council, Arts Council England, English Heritage, the Heritage Lottery Fund, Government Office for the South East and SEEDA.

34  For the closure of the Rotunda see Martin Easdown and Linda Sage, *The Demise of the Rotunda* (Folkestone: Marlinova, 2008); for Pleasureland see Steve Copnall, *Pleasureland Memories: A History of Southport's Amusement Park* (Wantage: Skelter Publishing, 2005) and, specifically on its closure, Anya Chapman and Duncan Light, this volume. Also closed during this period were Marvel's Amusement Park, Scarborough (1999); Dizzyland, Southend (2001); Merlin's Magic Land, St Ives (2003); and Once Upon a Time, Woolacombe (2006), as well as a large number of inland amusement and theme parks which suffered from the same problems.

35  Nick Laister, 'Proposed Heritage Amusement Park, Dreamland, Margate: Ride Availability, Concept Plan and Business Plan' (May 2007).

36  'Save Dreamland Campaign Launches Vision for World's First Heritage Amusement Park', Save Dreamland Campaign website (30 April 2007): http://www.joylandbooks.com/scenicrailway/news0704.htm accessed 27 June 2014.

37  *Thanet Extra* (4 May 2007).

38  http://www.thanet.gov.uk/council_democracy/consultation/dreamland_plan.aspx; for the Brief see http://thanet.gov.uk/media/2327714/21-Compressed-Dreamland-Brief-ADOPTED-BRIEF_FEB-08.pdf accessed 27 June 2014.

39  BBC News (11 April 2008): http://news.bbc.co.uk/1/hi/england/kent/7342647.stm accessed on 27 June 2014.
40  Prince's Regeneration Trust, 'Press Release' (24 June 2008).
41  Prince's Regeneration Trust, 'The Regeneration of the Scenic Railway and Dreamland Cinema, Margate' (July 2008).
42  Urban Panel, 'Margate Review Paper' (June 2009).
43  Rachel Cooke, 'Can Art Put New Heart into our Seaside Towns? *Observer* (16 August 2009). On this theme see Jason Wood, 'From Port to Resort: Art, Heritage and Identity in the Regeneration of Margate', in Peter Borsay and John K. Walton (eds), *Resorts and Ports: European Seaside Towns Since 1700* (Bristol: Channel View Publications, 2011), pp. 197–218.
44  'Sea Change Press Release', CABE website (16 November 2009).
45  *Kent on Sunday* (27 November 2011).
46  Earlier, in 2009, some of Sanger's original menagerie cages and mock abbey walls had been listed Grade II.
47  Thanet District Council, 'Press Release on CPO' (2 June 2011).
48  A design company run by Wayne, Gerardine and Jack Hemingway: http://houseofhemingway.co.uk/hemingway-design/

# 11 Designing the past

## Implications of a new vision for Dreamland

*Eleanor McGrath*

The ways in which most people think of the coastal amusement park, and more broadly the British seaside, are framed within individual and collective memory and accessed through representations in popular culture. This is now changing with the reopening of Dreamland in Margate as an experience-based visitor attraction which resurrects elements of a bygone world by offering to 'reach out' to people 'who cherish . . . the famous seaside amusement park'.[1] The construction of this new Dreamland may become a place where the past informs the present and the present electrifies and accentuates the past, whilst providing an intricate backdrop for the performance of memories and the creation of new cultural representations.

On 16 November 2013, the site of old Dreamland, closed to the public since 2005, was reopened to allow local people to meet the project team and understand the site's impending transformation. Open only for three hours, queues of the curious gathered to hear Nick Laister, Chair of The Dreamland Trust, make an impassioned speech as he cut the ribbon saying: 'It was time the people of Margate . . . take Dreamland back for themselves' (Figure 11.1).

Once on site a viewer could see the fragments of remaining structures – potent artefacts illustrating the history of the former amusement park and Margate's important role in the development of leisure time – from 'Lord' George Sanger's menagerie cages, to the 1920 Scenic Railway and modernist cinema building. These elements are now being reshaped and given new life, the culmination of a long battle, starting with the formation of the Save Dreamland Campaign and The Dreamland Trust, and with involvement of Thanet District Council and a commercial operator, to reclaim the site for use as a visitor attraction for Margate, rather than give it over for housing and retail development.[2] In addition to enhancing Margate's regeneration and restoring valuable cultural assets, the project gives recognition to a social history of leisure, a subject that has previously suffered from a value deficit within the academic and professional heritage sectors, as well as in public perception.[3]

Beyond the winding journey the project has taken to this point, and its significance as a site that can inform us about the heritage of British leisure and social change, the new Dreamland is of interest for the questions it raises within the field of heritage studies and beyond. The new Dreamland has been constructed

*Figure 11.1* Nick Laister reopens the Dreamland site, November 2013

Source: © The author

on the site of the old Dreamland using some historic rides from other places and adopting elements from the past with a blur of varied contextual references. It is a unique project that has been shaped by a team with the vision to take aspects of the old and make new, and an awareness of the need for commercial success to make the project viable. The growing audience will inform content and, driven by particular needs and desires, will further direct what the new Dreamland becomes.

The process of constructing and shaping a contemporary version of the past ultimately raises questions relating to authenticity, education and representation of memory. The distinctiveness of the project may draw the attention of critics, and although the heritage that Dreamland contains is linked to social history and leisure, the process of its creation links it to other heritage sites and visitor attractions across the world, including those that may convey more contentious histories. It is a project that could be framed within the wider academic spheres of heritage, historiography, design and philosophy, even if some may not consider it to be a heritage-focused endeavour.

This chapter sets out the contextual scene in which the new Dreamland project began, considering the implications of this in relation to the following themes: heritage, especially connotations of its meaning and use as a driver for regeneration; visitor experience driven by contemporary desires; authenticity as a sought-after concept, framed in opposition to the 'fake'; constructed architecture and experience-driven landscapes; memory, displacement and preferential narrative; motivation for attraction concepts; and design vision and editorial processes.

## Dreamland – Beneficiary of an evolved and accepting 'heritage'

In 2009, after many years of campaigning, the proposals for the new Dreamland were hailed as 'the world's first heritage amusement park', where visitors 'can step back to a time when rides were bigger and seaside amusements were truly a thrilling day out'.[4] The phrase 'heritage amusement park' could be said to be a slightly odd contradiction when framed within the well-trodden museum sector debates regarding the educational value of history and whether heritage interpretation distorts the past, weakening the delivery of potential historic narratives.[5]

As a reaction to the opening of attractions such as Jorvik in the 1980s, George MacDonald and Stephen Alsford argued that museums and heritage sites were increasingly having to adapt their offer when facing competition from theme parks, offering more experience-led attractions, leading to the question of just how 'fun' heritage should be?[6] Hewison critiqued these types of attractions on their emergence, observing that they sought to appeal, placed increasing revenue at the expense of education, and offered visitors emotional and evocative experiences over intellectual ones.[7] The provision of 'themed' or immersive experiences is something that continues to pervade and has been well documented elsewhere.[8] The recent reinterpretation of Britain's Second World War code-breaking centre at Bletchley Park[9] is just one current example whereby visitors can walk through the huts of the code breakers, with audio-visual effects and props used to give the impression that they have stepped back to war-time Britain, suggesting that the appeal of the set-dressed experience has not yet lost currency for many audiences. The eschewal by some critics of these experiences implies that they offer an inauthentic version of the past at odds with the official authorised version that is to be taught; an opinion that can be equally contested. Whatever the theoretical debates to the contrary, current heritage management practice beyond conservation is geared towards sustainability and an appealing visitor offer.

Marie Louise Stig Sørensen and John Carman note that heritage, rather than being appended to other disciplines, is now an explicit area of research in its own right, interdisciplinary in its nature and appropriation of methodologies. Heritage, they claim:

constitutes an influential force in society . . . expressed in the strong links between identity formation . . . [I]t may be approached as an object of study, or as a means of generating income, or as part of political action or sustainable development to engender community spirit . . . Some may see heritage as their inalienable right, whilst for others it may be a construct; yet others see it as timeless and belonging to all.[10]

Clearly, therefore, the word heritage is multifaceted – and powerful – encapsulating levels of meanings and interpretations based on the ideas of the individual, the collective, the national and the transnational.[11]

Heritage as a term provides further complication as it is often adopted by those engaged in statutory conservation.[12] In the context of amusement parks, beyond the listing of the Dreamland's Scenic Railway and the Water Chute in Hull's East Park, there are complexities regarding the designation of historic rides.[13] At other parks, such as Blackpool Pleasure Beach, there has been a detectable reluctance to enter into a formal protected status or adherence to a collecting or de-accession policy, perhaps because the word heritage tends to be loaded with protectionism, conservation and designated grading systems, which when applied may entrap what would otherwise be dynamic and changing spaces.

Historically amusement parks were places in flux, with attractions shifting between seasons to draw visitors back and meet changing expectations. Whilst, for example, Blackpool's Big Dipper and Grand National are perhaps less at risk, a move to apply heritage protection to smaller, more portable elements may alarm those charged with operating viable visitor attractions for fear the heritage emphasis aligns them more closely with outdoor museums, such as Beamish and the Weald and Downland Open-air Museum, where policies are in place to safeguard their collections.

Based on the idea that people respond differently to various heritages and the term's wider connotations, it is understandable that the inclusion of the word 'heritage' in the new Dreamland's initial overarching vision as the 'world's first heritage amusement park' might have unintentionally weighed down the project concept. By 2013, the tagline had changed. The heritage word was dropped and instead the new Dreamland promised to be 'the world's first amusement park of thrilling historic rides,[14] communicating both vibrancy and clarity with regards to what the park would contain. In the latest iteration, the language has changed again to the 'UK's original pleasure park: re-imagined',[15] conveying both the original and authentic, alongside transformative and contemporary updating.

So, why did the new Dreamland need to be underpinned by heritage and could it have proceeded without the heritage sentiment? Heritage as a concept can arguably be seen as the element of Dreamland that made it a viable project. The listing of the Scenic Railway had a role in protecting the site and surrounding land from aspects of development. Going further, the viability of installing new rides, of a quality standard similar to contemporary theme parks, would have been too risky and costly; historic rides could be donated or purchased for less money, whilst avoiding the attraction becoming stifled by generic and contemporary portable rides which can be found frequently at other locations. The historic rides would also help ensure uniqueness, but as with all conservation projects they could not be considered a cheap option. The heritage element unlocked new and additional funding sources. For example, a grant commitment of nearly six million pounds was secured as a result of a successful application to the Heritage Lottery Fund (HLF).[16] Commitment of this level of funding not only demonstrates the learning and participation opportunities for local people within the project but also the heritage value of the assets and the importance of the redevelopment of the site for Margate. Rather than heritage entailing a static and protection-based outlook, it is more often than not a dynamic driver for change, positively contributing to the conception and delivery of sustainable regeneration projects.[17]

As Laurajane Smith has argued, English attitudes have been commonly driven by the need for perceived authenticity and authorised value systems set by a small set of organisations within the heritage sector.[18] But as the HLF investment demonstrates, at Dreamland and other deprived resort towns, the value of Britain's seaside history and associated memories and architecture, once dismissed as vernacular or 'low' in contrast to the 'high' histories of war, monarchy and architectural classification, is now being embraced, offering the potential to engage new audiences and allow for lesser-known histories to be interpreted.[19] It could be argued that heritage has been reclaimed from conservative foundations and is now an open and accepting concept. HLF has an open approach to heritage; it is inclusive and not defined by prerequisite knowledge and limiting frameworks, as Robert Hewison and John Holden succinctly remark:

> [HLF] remains steadfast in the refusal to give a potentially limiting definition to the word 'heritage', allowing the word to be defined by the practices it chooses to support in response to the applications it receives. The philosophical approach to heritage in HLF is therefore different to that of some other heritage organisations who use their expert knowledge to identify, manage and advise on what is important on behalf of society.[20]

Dreamland is important to the people of Thanet, Kent and beyond. The richness of the heritage within the site is not false or inflated; it is organic, well-remembered and supported by strong local memory and archive material. This was clearly recognised, along with regeneration opportunities for Margate, by HLF in their decision to support the project along with other funders.

So heritage, complicated word that it is, will underpin what the new Dreamland will become, ensuring the past is drawn upon but layered with contemporary ideas of what the experience should feel like and used as a catalyst for positive community benefit. The social element is potentially the strength of the project, placing the proposals in the heart of the local community in Margate. At the grand opening of the Scenic Railway in 1920, the proprietor John Henry Iles declared that 'Dreamland would have an important bearing on the future prosperity of Margate'.[21] Arguably the town cannot easily dissociate its local identity from Dreamland itself.

## Visitor desire and contemporary appeal

If the new Dreamland is to succeed it will need to appeal to new audiences, some of whom may not understand the idea of an amusement park, perhaps never having experienced one first hand. New audiences are likely to be more familiar with other modes of leisure attraction, be they a shopping centre, art gallery, theme park, heritage site or museum.[22] Sociologists such as George Ritzer and Alan Bryman have argued that the contemporary visitor has been exposed to and participated in mass-cultural experiences which have been shaped by global corporations such as Disney and McDonalds. Audiences are drawn to appealing destinations that Ritzer terms 'cathedrals of consumption' whilst Bryman has argued that Disney, in its offering of efficient models of fantasy and service standards, has directly impacted on the shape of visitor attraction types which have to appeal to visitors with a preference for the expected, sanitised and thematic experience.[23] The honed and homogenised experience offers familiarity and enforces expectations that extend to other public spaces and attractions whether corporate in nature or not. Following these arguments, the new Dreamland as a visitor attraction must appeal to a mass-society which has been described as increasingly 'consumerist', 'nostalgic' and 'escapist'.[24]

Written into the strategy for the new Dreamland was that it must be a viable and successful visitor attraction with wide appeal, and operating on a two-layered approach: the first appealing to the general visitor who would enjoy the attractions and food offer (these may include families, the 'arty daytripper', those who appreciate the history of popular British culture and the local community); the second acknowledging the seasonal difficulties that coastal resorts face by marketing itself as a year-round venue for large events (such as music festivals, vintage and collector fairs, classic car rallies), as well as for private or corporate hire.[25] The new Dreamland comes with strong branding from the styling of staff uniforms to the opportunity to buy merchandise stamped with the park's logo. There is even the opportunity to buy 'Scenic salvage' goods, such as candle-holders and key-rings, made from the old timbers of the Scenic Railway that could not be reused in the ride's restoration (Figure 11.2).

Whilst the new Dreamland will not be marketed as a theme park it may possess features that align it with these types of attractions in terms of theming, landscaping, architectural conflations, adoption of popular motifs, constructed narratives and boundaries creating an interior and exterior world. Framing new Dreamland as an amusement or pleasure park acknowledges the site's important past but in some ways also provides it with its particular 'theme' or narrative. Theme parks have been considered as derisory or low culture; the rejection of the term displays a message that the attraction on offer is not fake or shallow; it is authentic and organic, as opposed to a contrived experience.

## Proclamations of the authentic experience

In the context of heritage sites the false versus the authentic is not something that can be universally applied, indeed there are spectrums of both elements present in many types of attractions that are underpinned by the past. Nor can it confidently

*Figure 11.2* A piece of 'Scenic salvage'. Visitors to Dreamland can purchase key-rings made from the old timbers of the Scenic Railway

Source: © The author

be said whether visitors tend to prefer the attraction that proclaims to be authentic, or the one that offers an immersive and escapist experience. Whilst some visitors may be drawn to constructed environments, for others they may actively reject experiences perceived as inauthentic; hence 'authenticity' becomes a marketing tool and something that brands seek in a world with mega corporations and high street homogeneity.[26]

At certain heritage sites notions of authenticity are what underpin their authority, and hence dictate their visitor numbers. In their studies of America's Colonial Williamsburg, Eric Gable and Richard Handler examined how staff differentiated themselves from other constructed landscapes such as Disneyland. Staff argued that Colonial Williamsburg was a serious educational institution, founded on research. Where the public, perhaps not in the majority, might be sceptical about curatorial decisions or site representations, the idea of authenticity holds resilience as it is perceived to be under threat and also through Colonial Williamsburg's careful admission of where incorrect decisions have been made in the past, further enhancing the site's image of being underpinned by validated historiographical presentation and interpretation.[27]

Walter Benjamin discussed the impact of modernity on the authentic object, conscious that widespread adoption of film and photography could replicate images with such ease that they could jeopardise the authority of the authentic object and the historic testimony held within it. Tony Bennett has argued that the object itself did not directly possess authority with absolute autonomy, but rather it fell into a representational machine when framed within museum contexts; it was the way in which an object was framed that loaded it with meaning.[28] Moving these opinions to the present day, Martin Hall has argued that an 'experimental complex' exists where the physical, and perhaps deemed authentic, object has been appropriated into contexts including and beyond museum and heritage locations. In an interrogation of a multitude of sites, he observes that these simulated experiences depend on 're-injecting realness', and that through inclusion of authentic cultural objects they provide an anchor to the simulacrum or the created space.[29]

Throughout the twentieth century a number of theorists have interrogated the boundaries between the 'real' and the 'fake'; Marc Augé even arguing that the abundant presence of the past could, through the repeated recycling of images, overwhelm the present day, impeding the ability to distinguish between original elements and later representations.[30] Within the context of a new Dreamland, a future could exist whereby a visitor may experience the site, considering it a full replica, or conversely a site without history if the boundaries of intervention are not expressed. Whilst lines can be blurred between 'real' and 'fake' it would be simplistic to treat them as polarised, or assert that one is morally better than the other.[31] As Umberto Eco and Jean Baudrillard both explore, the inauthentic, or false, is not always constructed with intent to deceive. Integrity to how something has been created is key; something that is a reproduction, and thus materially 'totally fake' can be considered more authentic as it is not actively seeking to deceive. Baudrillard argues that constructed or suspended spaces such as Disneyland have integrity, as they can be judged as 'real fakes' and therefore true, as opposed to false, as they do not seek to directly replicate and mislead the viewer.[32] Dreamland, as an immersive visitor attraction, will rely on interpretation which draws out the historic underpinning alongside artistic response and commercial interventions.

## Architecture as a dangerous backdrop

There is certainly danger that physical alterations to contextually correct spaces may confuse the viewer in their ability to differentiate between physical remnants and later interventions. Intervention could imply total reconstruction, tweaks to visitor presentation, or the conflation of new architecture to imitate existing structures. These arguments are particularly relevant to the new Dreamland as a contextual site with residual heritage assets being integrated within the construction of a modern visitor attraction.

Colonial Williamsburg provides a useful case study. Bankrolled by John. D. Rockefeller in 1927, the ex-colonial capital of Virginia began a transformation backwards into an example of an eighteenth-century town. By purchasing building by building, and stripping back and restoring each one, a firm of Boston architects recreated the town as 'composite representation of the original forms of a number of buildings and areas known, or believed to have existed between 1699 and 1840'.[33] The site today bears little trace of the town it was before 1927. The buildings are marked as numbered attractions, with discreet signage unapologetically noting the date of restoration and thanking the individual donors who helped pay for their transformation.

The most disconcerting experience is perhaps upon entrance to the star-billed Governor's Palace, where costumed tour guides offer an experience akin to certain National Trust properties, whereby artefacts, interior décor and the lives of the family who lived there are provided in great detail. The guides and interpretation, however, omitted the story of the building's 1930s demolition to allow for complete reconstruction (on top of surviving sixteenth-century wine cellars) to transport the house to the mid eighteenth century, complete with period interiors and paint designs based on, and subsequently adjusted by, research.[34] To the unquestioning eye, this beautiful building, rich in acquired antiques and lavish decorations, would not only appeal to certain ideas or desires for a heritage-based experience, it would be accepted as completely authentic. Whilst the story of how Colonial Williamsburg came to be is arguably just as interesting as other aspects of the site's history, the marketing of the site as an authentic experience plays down the level of intervention undertaken to transform it into a time-capsuled vista.

The way that the past is represented at heritage attractions is not dissimilar to the way that theme parks are constructed: Lowenthal, for example, has observed that theme park landscapes and landscapes of the past are both 'an artifice, an invention, a construct, an illusion'.[35] It is not just direct reconstructions or new buildings, but the placement of genuine artefacts to provide appeal. John K. Walton, in his analysis of Beamish, noted that whilst the buildings present at the site are very much authentic, they have been decontextualised and placed within a 'fake urban environment'. Visitors can go to the Sun Inn or to a sweet factory on

Beamish's Market Town Street, allowing them to experience and consume ideas which hold attraction.[36] This idea directly correlates to a new Dreamland where portable rides are being re-contextualised and pathways and landscaping are being created within a new environment; all to be delivered in response to visitor expectation and the desire to perform experience-based activities they seek out.

When done convincingly, it seems that enjoyment of replicas is popular whether in theme park contexts or heritage ones. Anton Clavé has suggested that people are aware of unreal aspects of the visitor experiences but argues that for many 'clichéd statements of culture are what they seek'.[37] Dependent on context, the creation of representations informed by jumbled epochs, pop culture reference and the viewpoint of the individuals constructing the space can be dangerous. The Lost World, an African theme park, sought to represent images of the entire continent in one place; however the images represented were felt to correlate to a colonial viewpoint, further entrenching African stereotypes, and the park closed not long after opening in 1992.[38]

The new Dreamland will have to impress and inspire visitors, and draw them into a cohesive landscape through the appropriation of cultural images, references and style. Wayne Hemingway highlighted the importance of visual impact for the new Dreamland, acknowledging that it has to look good and provide photographic opportunities, from seaside cut-outs to old ride cars, in every nook and cranny.[39] Arguably, this representation of fantasy has gone further with the rise of social media, with Instagram in particular promoting self-documentation and hash-tagging of experiences. Photographic self-documentation extends to monumental and civic art, and to sites of contention or memorialisation[40] informed by the landscape or architectural backdrops created.

The reasons for visiting and self-documentation may be driven by tourism and associated behaviours or an emotional need to connect with a past, or undertake commemorative activity. David Uzzell has observed the different ways in which places with emotional memory can be interpreted when he considers 'heritage that hurts', or 'HOT interpretation'.[41] Whilst Dreamland is not a site of human tragedy, it is a site with strong local memory still in existence and hence may evoke emotional responses. The site, although different to an individual's memory, may be so skilfully constructed that the motifs and ideas within it have the potential to displace memory, creating 'social amnesia' where people believe they partially remember by experiencing the recreation.[42] A concern could be that the popularity of the new Dreamland unintentionally results in displacement of seaside culture from elsewhere, presenting us with a perfect example of a seaside with the photographic backdrop people seek. Certainly, this is possible, as Hemingway feels that the uniqueness of the attraction could make it a global destination, drawing not just UK visitors but those from abroad.[43]

## Preferential narratives and selective memories

Beyond architectural constructions, the narratives displayed at heritage sites are also selective. Cornelius Holtorf has argued that people are not just seeking an authentic experience but sometimes an emotional connection, which may in turn displace factual information.[44] Through promotion of shared and living memory and the emphasis on particular eras and inspirations from the twentieth century, the new Dreamland could be argued to be promoting an attraction based on nostalgia. Although the vision is clearly for a contemporary, or twenty-first-century experience, it is somewhat underpinned by memories and ideas that link to a previous time, or extend more broadly to encompass the character of the 'Great British seaside' experience[45] and with the 'baby-boomer' generation in mind as one of the possible audiences.[46]

The desire to escape modern life and seek experiences based within the past could be argued to be driven by nostalgic impulse, if not for a time within one's own lifetime, for a preferential time or civilisation. Nostalgia has been a source of anxiety since its classification as a disease in the seventeenth century, to Robert Hewison cautioning that nostalgia is a symptom of the postmodern world where our reliance on the past is unhealthy.[47] Driven by loss, or absence, nostalgia comprises an editing process where selective memories may displace aspects of the past that are less favourable. The concern of some critics is that progress is hindered and representations of what is perceived to be authentic are continually recycled, while participants may intentionally reconstruct aspects of the past they know to be false, driven by preference. David Lowenthal terms this 'anti-heritage'.[48]

A visitor services manager at Dickens World in Kent made the point that entertainment is the primary aim as a true account may depress visitors: 'We're not going to have starving babies crawling around on cobblestones'.[49] Motivation beyond encouragement of entertainment could be based on intentional avoidance to address difficult histories. It has been argued that Colonial Williamsburg, whilst acknowledging elements of slavery and attitudes towards mixed-race relationships within its history, downplays this narrative, readdressing it through specialist tours outside of the core visitor experience.[50] Colonial Williamsburg places itself as a site of freedom, focusing instead on American narratives of independence from the British, and hence is intertwined with ideas relating to nationhood, and the nostalgia for a 'pure' point in time.

The new Dreamland may also draw on popular epochs within the site's history. The 1960s are likely to be more popular in the mind than say the 1990s or the Edwardian eras, with the latter being too remote and the former too recent and indicative of the site in a period of decline, or an undesired seediness or danger. The 1960s was a period when rides were exciting, pop culture in a postwar climate was growing and the mods versus rocker battles on Margate beach on Whitsun 1964 became legend. Hemingway cites Margate as being 'steeped in British youth culture, and British youth culture had a big influence on the world', so to not include it in ideas relating to design at the new Dreamland would be 'remiss'.[51] Whilst those with living memory may still hold interest in and wish to participate in scooter rallies, this may not always have fashionable currency, and perhaps the presentation of youth culture should contextualise the clashes on Margate beach as an important part of local history, rather than simply as a motif to inform visual appeal at Dreamland.

Although the new Dreamland will draw on local memories and histories, as an attraction it may avoid being consumed by the nostalgia trap through the assurances that contemporary vision and design are in equilibrium within the approach. Hemingway clearly understands the strong local connections and memories present, describing Dreamland 'as part of the DNA of Margate'. He makes clear that the vision for the site can draw from the past without 'bringing it back in its purest form . . . it has to be relevant for people who are not really interested in looking back, but just want something good'.[52] Indeed, it is an important point that nostalgic impulse is not universal or endemic across visitor groups. The new Dreamland will need to appeal to those who wish to remember, and those who are motivated more by contemporary experiences which are relevant to them.

## Motivation for conception and design vision

When Lowenthal remarked that theme park landscapes were informed by images inspired by 'wishful and wilful geographies of the mind',[53] he could also have been speaking of the open-air museum, amusement park and heritage site. Explorations of the authentic and ideas of constructed spaces indicate common threads across visitor attraction models; their subtleties are perhaps best differentiated by both the motivation behind their conception and the editors involved in their creation.

At Colonial Williamsburg, the concept was born out of preservation, with the Revd Dr W. A. R. Goodwin inspired by the seventeenth-century parish church within his care, extending his vision to preserve Williamsburg as an 'introduction to colonial America' and heart of the nation's history.[54] While at Beamish, founder Frank Atkinson was also motivated by preservation, but for social histories relating to everyday life. He felt strongly that museums had a role beyond the curatorial expert, deploying the phrase 'enrichment through enjoyment' to offer the public opportunities to relate to the past in an engaging and pleasant context.[55] Walt Disney, although inspired by the past through his idealised vision of his own upbringing in turn-of-the-century Missouri, had entertainment at heart, with educational literature and history underpinning the references adopted, especially in the creation of Main Street USA. Whilst it is possible to learn at a Disney park, that was not the main aim. Nostalgia, however, had an element to play as Disney remarked as early as the 1940s: 'It's a shame people come to Hollywood and there's nothing to see . . . they expect to see glamour and movie stars'.[56] The new form of Disney's park was to build upon ideas of what people desired but in a safe environment, away from the 'tough-looking people' and illicit and dangerous world of the carnival and American amusement park.[57]

Beyond their conception, sites of attraction are further created and manipulated by an editorial process. We may differentiate with job titles, with preservation architects and curators at Williamsburg, curators and interpretation masterplanners at Beamish, imagineers at Disney or creative design consultants at Dreamland, but all are undertaking a process of discarding and active selection to inform a unique vision and shape for what each attraction looks like and the messages conveyed, sometimes offering mediation between curatorial control and visitor engagement.

Disney's imagineers do not seek to replicate the past or national landscapes with EPCOT's World Showcase, rather they work to a concept of what one of them has termed 'Disney realism', a concept that is 'sort of utopian in nature, where we carefully programme out all of the negative, unwanted elements, and programme in all of the positive'.[58] Countering this idea, the artist Banksy opened his artistic concept 'Dismaland' in the seaside town of Weston-super-Mare, Somerset, in August 2015, offering a family theme park experience, 'unsuitable for small children', providing a subversive and anarchic experiences directly inspired by amusement landscapes and Disney theme parks, expressing the need to expand upon the conventional thematic experiences society is exposed to.[59]

Whilst the 'theme' for the new Dreamland is rather self-referential and the site contextually 'real', the visual messages and hence the visitor experience would not emerge successfully without the appointment of architects, and a designer with vision. Deyan Sudjic has observed that 'design has been used to engineer desire since it first emerged as a distinct profession'. Objects are not passive or innocent; they are loaded with subtle references indicated by shape, colour choice and material.[60] The new Dreamland is not only formed on context, perceived memory and the built landscape, but also to the portable physical 'things' that populate it, available to interact with, or available to purchase as souvenirs.

To avoid becoming a theme park space where time is conflated and visitors are stuck in a purely nostalgic 'good old days', pre-1969 landscape, the skill lies very much with Hemingway as a designer. His approach provides reassurance. At the 'Calling All Creative Minds' event held in April 2013, Hemingway challenged the community in Margate to help bring Dreamland to life. This ensured ownership as well as creative opportunities. Also, given the limited budget, advice and support from local businesses was going to be important to ensure longer-term resilience either through donation, collaborative-working, events or sponsorship.

HemingwayDesign actively seek to avoid terms such as 'retro' and 'nostalgia' assimilating with their design brand. Hemingway cautioned that their design process is to only 'select things which have relevance today . . . if you rely solely on heritage and nostalgia, it is a dangerous game to play'.[61] The new Dreamland will not be specific to any particular decade. Hemingway has also said their approach is about 'the beauty of the past being kept alive for the future'[62] such as for mid-century furniture maker G-Plan, where they have taken a classic design but put a contemporary stamp on it. A Hemingway design exists with its own distinct colour palette and design, standing alone as an independent viewpoint. The fresh approach helps avoid the idea that things inspired by the past have to be prefixed by 'repro-' or 'retro-' comprising pale imitations of the original.[63] In many instances Hemingway possesses an editorial eye that enlivens, or even improves many of the inspirations from the past.

Ensuring the site has wide appeal has meant consideration of how the new Dreamland will counter the amusement park's traditional seasonal operating hours, avoiding also the concerns of seasonal employment in resort communities. The larger aim is to create a cultural and creative event space that can host events all year round.

If Dreamland is to be a contemporary attraction, a way to temper past influences will be through inclusion of artistic or creative response. At the Calling All Creative Minds event, Hemingway acknowledged that the creative response could also produce an exciting aesthetic finish through the upcycling of materials and found objects, further counteracting the small budgets involved to produce a quality attraction. He observed again that the upcycling process should not be undertaken in 'a twee Cath Kidston way', but a 'subversive' one, further emphasising the need for the creative response to be strong, independent and shunning typical vintage pastiches. This has been reflected in the upcycling of traditional rides, with the 1970s dodgem cars stencilled with famous stars and offering visitors the opportunity to ride the 'Counterculture' or 'Kiss-me-Quick' Caterpillars.

Margate and Dreamland possess character, heritage and a unique identity in abundance.[64] Hemingway made a powerful statement when he declared 'the aesthetics of this town have wide appeal, but it has got to be done right'.[65] Certainly, in his role as designer, Hemingway is not seeking to displace the elements that survive, or simply use Dreamland as a vehicle for purely his own aesthetics or self-promotion. The project will continue to be run in conjunction with the community and will use their feedback and involvement to create an integrated and strengthened result.

## Conclusion

The new Dreamland will be a unique visitor attraction embracing an amusement park heritage and perpetuating a tradition of continual change in response to visitor demand. By using heritage as the point of departure to layer new elements on top of a contextual site, the new Dreamland is aligned with other modes of attraction such as the theme park, open-air museum and wider constructed landscapes. Concerns relating to displacement and notions of the authentic can be resolved through narrative that tells the whole story, ensuring a visitor twenty years from now will be able to distinguish between the project works and original heritage assets, as well as realising the former context of imported historic rides. It is crucial, too, that Dreamland's less marketable histories are not forgotten, including the story of the Save Dreamland Campaign.

The new Dreamland takes local memories, popular motifs, enthusiast knowledge and fragments of amusement park history and heritage and places them back into mainstream culture. For this is a project about place-making, and for Dreamland to go forward and be successful it has to be bold and forward thinking, re-inventing its traditions but not being constrained by them. The challenge remains to ensure the heritage underpinning the concept remains undiluted. As Phase One of the project opened for business on 19 June 2015, the inclusion of temporary rides to bolster the rostra of attractions for visitors did not pass unnoticed. The new Dreamland's initial driving force was that of preservation of historic rides, like the beautifully restored 1950s Hurricane Jets (Figure 11.3) and the associated contextual buildings. The operators of the park will need to retain this emphasis on heritage significance and avoid any slippage back to a precarious

*Figure 11.3* The restored Hurricane Jets in front of the Scenic Railway, July 2015

Source: © The author

and typical seaside attraction which leans too heavily on contemporary portable and standardised funfair attractions that can be found anywhere. Of course, the star of the show is the Scenic Railway (reopened in October 2015 after lengthy restoration), and the park does feature 'rides from every decade from the 1920s to today'.[66] The park also pays homage to its history with timelines, archival imagery and interpretation at the entrance of each historic ride.

There is no denying that this has been, and continues to be, a complex project. With further phases to come, the park will face financial constraints when compared to attractions that might have larger budgets or a sole commercial interest. The new Dreamland will continue to require creative thinking, adoption of social media and support from others in the amusement sector such as Blackpool Pleasure Beach, which may, in addition to rides already donated, provide further assistance and inspiration.[67] Certainly, Blackpool Pleasure Beach, and attractions such as Carters Steam Fair and Dingles Fairground Heritage Centre, demonstrate that vintage rides are both thrilling and have market appeal. The key to success is a strong project team, supportive Trust, involved community and an unwavering commitment to an overall vision for the site.

Above all else, the new Dreamland must be re-integrated within Margate. Disney's motivation for creating his parks was to keep elements he deemed unsavoury out, with the modern day Disney parks critiqued for their prohibitive ticketing prices.[68] In April 2013, the project team indicated that the new Dreamland would be without a pay barrier, with the landscaping and site accessible to all, and with admissions to each attraction being on a pay-as-you-go basis or possibly via a wristband.[69] By the time the park opened, business planning appears to have changed the approach, with the installation of a pay barrier of £17.95 in place for ride access, although several features such as the ballroom and roller-rink are free. Edward Relph has remarked that 'an authentic sense of place is above all that of being inside and belonging to your place as both an individual and as a member of the community, and to know this without reflecting on it'.[70] The absence of pay barrier enforces Relph's ideas of authentic and integrated spaces, further facilitating inclusivity, breaking down physical boundaries of otherness as seen in theme park contexts and ensuring great numbers of people access the site. It is perhaps unfortunate that the pay barrier policy has been implemented. Having said that, Thanet residents can gain discounted access and the reopening of Dreamland has undoubtedly provided a significant boost to the local economy and provision of jobs.

As a visitor model the new Dreamland offers new life for historic rides, protection for designated heritage features and a visitor attraction which will further support Margate's wider cultural and regeneration strategies.[71] It is a small panacea in a larger world, but the potential popularity of it may inspire more people to visit seaside towns and perhaps gain an understanding of the British coast's importance to social and leisure history. The approach at the new Dreamland has challenged traditional notions about heritage. If heritage is not just about preservation, and involves the active response to societal need or desire, placing the past in the present, then it will be important to acknowledge the park's constructed nature, the choices which have been made, and the people who were involved in a lengthy campaign to save it.

## Notes

1  The Dreamland Trust, 'New Dreamland: A Vision for Dreamland Margate' (2013): http://www.dreamlandmargate.com/new_dreamland.html accessed 2 July 2014.

2  See Nick Laister, this volume. For the site's history see Nick Evans, *Dreamland Revived: The Story of Margate's Famous Amusement Park*, rev. edn (Whitstable: Bygone Publishing, 2014).

3  See, for example, John K. Walton and Jason Wood, 'World Heritage Seaside', *British Archaeology* 90 (2006), pp. 10–15, which makes an argument for Blackpool's nomination as a UNESCO World Heritage Site. On the theme of the placement of the popular in history see Emma Griffin, 'Popular Culture in Industrialising England', *The Historical Journal* 45, 3 (2002) pp. 619–35.

4  The Dreamland Trust, 'Our Vision' (2009): http://www.joylandbooks.com/scenicrailway/heritageamusementparkvision.htm accessed 30 September 2014. This has been archived online and can still be accessed, although now displaced by revised vision documents.

5  On this theme see Eilean Hooper-Greenhill, *The Educational Role of the Museum* (London: Routledge, 1994); also Rodney Harrison, *Heritage: Critical Approaches* (London: Routledge, 2012), pp. 109–10, for a summary of debates on educational versus curatorial roles and the economic motivations of museum and heritage sites.

6  George MacDonald and Stephen Alsford, 'Museums and Theme Parks: Worlds in Collision?', *Museum Management and Curatorship* 14, 2 (1995), pp. 129–47.

7  Robert Hewison, *The Heritage Industry: Britain in a Climate of Decline* (London: Methuen, 1987), p. 21.

8  See, for example, Cornelius Holtorf, *From Stonehenge to Las Vegas: Archaeology as Popular Culture* (Walnut Creek: Alta Mira Press 2005).

9  Based on a site visit made by the author in May 2014, prior to opening.

10 Marie Louise Stig Sørensen and John Carman, 'Introduction: Making the Means Transparent: Reasons and Reflections', in Marie Louise Stig Sørensen and John Carman (eds), *Heritage Studies: Methods and Approaches* (London: Routledge, 2009), p. 3.

11 Sharon J. MacDonald, 'Museums, National, Postnational and Transcultural Identities', *Museum and Society* 1, 1 (2003), pp. 1–16.

12 For a summary of the legislative and organisational framework within the heritage sector in England see Simon Thurley, *Men from the Ministry* (London: Yale University Press, 2013).

13 See Allan Brodie and Roger Bowdler, this volume.

14 HemingwayDesign, 'HemingwayDesign Presents a New Vision for Dreamland Margate' (2013): http://www.hemingwaydesign.co.uk/projects/design/dreamland-margate; and Dreamland Trust, *Things to Come* (2013): http://www.dreamlandmargate.com/things_to_come_page.html accessed 30 June 2014.

15 http://www.dreamland.co.uk/ accessed 14 June 2015.

16 Heritage Lottery Fund, 'Heritage Lottery Fund Awards Dreamland £3m' (2011): http://www.hlf.org.uk/news/Pages/HeritageLotteryFundawardsDreamland3m.aspx#.U15L29yprwI accessed 30 June 2014; https://www.hlf.org.uk/about-us/media-centre/press-releases/uk%E2%80%99s-original-pleasure-park-dreamland-margate-re-opens-public#.VZQEQWYtDcs accessed 19 June 2015.

17 On this theme see Heritage Lottery Fund, *New Ideas Need Old Buildings* (London: Heritage Lottery Fund, 2013): available at http://www.hlf.org.uk/aboutus/howwework/Documents/NIOB_2013.pdf.

18 Laurajane Smith, *Uses of Heritage* (London: Routledge, 2006), p. 12.

19 The largest HLF grant in this context has been awarded to Blackpool Council for a major project to develop a museum to tell the 'extraordinary story of the world's first working-class seaside resort, celebrating its contribution to British and Western popular culture':

Heritage Lottery Fund, 'Blackpool Moves One Step Closer to £20 million Museum Dream' (2014): http://www.hlf.org.uk/about-us/media-centre/press-releases/blackpool-moves-one-step-closer-%C2%A320million-museum-dream accessed 3 August 2015.

20 Robert Hewison and John Holden, *Challenge and Change: HLF and Cultural Value* (London: Demos, 2004), p. 21.

21 Evans, *Dreamland Revived*, p. 15.

22 The differences between these types of attractions, and what distinguishes them from the amusement park, is contrasted and explored by Gary S. Cross and John K. Walton, *The Playful Crowd: Pleasure Places in the Twentieth Century* (New York: Columbia University Press, 2005).

23 George Ritzer, *The McDonaldization of Society: An Investigation into the Changing Character of Contemporary Social Life* (London: Pine Gorge, 1996); Alan Bryman, *The Disneyization of Society* (London: Sage, 2004); see also S. Anton Clavé, *The Global Theme Park Industry* (Wallingford: CABI, 2007), p. 167.

24 Ritzer, *The McDonaldization of Society*; MacDonald and Alsford, 'Museums and Theme Parks', p. 132; Scott A. Lukas, *Theme Park* (London: Reaktion, 2008), p. 7.

25 HemingwayDesign and The Dreamland Trust, 'What is Dreamland?' (2013): http://www.hemingwaydesign.co.uk/projects/design/dreamland-margate accessed 30 June 2014.

26 On this theme see David Lewis and Darren Bridger, *The Soul of the New Consumer: Authenticity – What We Buy and Why in the New Economy* (Boston: Nicholas Brealey Publishing, 2001), p. 3; also Fredric Jameson, *The Cultural Logic of Late Capitalism* (Durham: Duke University Press, 1991).

27 Eric Gable and Richard Handler, 'After Authenticity at an American Heritage Site', *American Anthropologist*, new ser., 98, 3 (September 1996), pp. 568–78.

28 Walter Benjamin, *Illuminations: The Work of Art in the Age of Mechanical Reproduction* (London: Pimlico, 1999); Tony Bennett, *The Birth of the Museum: History, Theory, Politics* (London: Routledge, 1995).

29 Martin Hall, 'The Reappearance of the Authentic', in Ivan Karp, Corinne A. Kratz, Lynn Szwaja, Tomas Ybarra-Frausto (eds), *Museum Frictions: Public Cultures and Global Transformations* (Durham: Duke University Press, 2006), pp. 93–4.

30 Marc Augé, *Non-Places: Introduction to an Anthropology of Supermodernity* (London and New York: Verso, 1995), pp.75–8.

31 See Edward Relph, *Place and Placelessness* (London: Pion Press, 1976), p. 80 for discussion of Nietzsche and Heidegger's morality regarding the inauthentic.

32 Umberto Eco, *Travels in Hyperreality*, trans. William Weaver (San Diego: Harcourt, 1986); Jean Baudrillard, *America*, trans. Chris Turner (London: Verso, 1988).

33 George Humphrey Yetter, *Williamsburg Before and After: The Rebirth of Virginia's Colonial Capital* (Williamsburg: Colonial Williamsburg Foundation, 1988), p. 60.

34 For details regarding the re-presentation of the Governor's Palace see Eric Gable, Richard Handler and Anna Lawson, 'On the Uses of Relativism: Fact, Conjecture and Black and White Histories at Colonial Williamsburg', *American Ethnologist* 19, 4 (November 1992), pp. 791–805.

35 David Lowenthal, 'The Past as a Theme Park', in Terence Young and Robert Riley (eds), *Theme Park Landscapes: Antecedents and Variations* (Washington: Dumbarton Oaks, 2002), p. 14.

36 Cross and Walton, *The Playful Crowd*, p. 229.

37 Clavé, *The Global Theme Park Industry*, pp. 218, 263.

38 Jeanne van Eeden, 'The Colonial Gaze: Imperialism, Myths, and South African Popular Culture', *Design Issues* 20, 2 (2004), pp. 18–33.

39 Wayne Hemingway speaking at the 'Calling All Creative Minds' event held at Turner Contemporary (April 2013). HemingwayDesign were appointed as the creative design consultants for the new Dreamland project in November 2012: http://houseofhemingway.co.uk/hemingway-design/projects/dreamland/. Coming from Morecambe, another

seaside town, Wayne and Gerardine Hemingway have an appreciation of British coastal resorts and are known for their inspiration, rather than appropriation, of past design to inform contemporary and exciting work.

40  John Lennon and Malcolm Foley, *Dark Tourism: The Attraction of Death and Disaster* (London: Continuum, 2000).

41  David Uzzell, 'Heritage that Hurts: Interpretation in a Post-modern World', in David Uzzell and Roy Ballantyne (eds), *Contemporary Issues in Heritage and Environmental Interpretation: Problems and Prospects* (London: The Stationary Office, 1998), p. 152–71.

42  Kevin Archer, 'The Limits to the Imagineered City: Sociospatial Polarisation in Orlando', *Economic Geography* 73 (1997), p. 336.

43  Interview with Wayne Hemingway (21 October 2013).

44  Holtorf, *From Stonehenge to Las Vegas*.

45  Dreamland Trust and Margate Renewal Partnership, 'Dreamland Margate: Sea Change Application' (2009): http://www.dreamland.org.uk/dreamland_vision.pdf accessed 30 June 2014.

46  Locum Consulting, 'Business and Audience Development Plan: Dreamland' (2009), p. 39: http://www.dreamlandmargate.com/downloads/3_Dreamland_Margate_SC_Dreamland_Business_Plan.pdf accessed 30 June 2014.

47  Svetlana Boym, *The Future of Nostalgia* (New York: Basic Books, 2001); Hewison, *The Heritage Industry*, p. 132.

48  Susan Stewart, *On Longing: Narrative of the Miniature, the Gigantic, the Souvenir, the Collection* (Baltimore: John Hopkins University Press, 1984), p. 23; David Lowenthal, *The Heritage Crusade and the Spoils of History* (Cambridge: Cambridge University Press, 1998), p. 100.

49  Quoted by Lukas, *Theme Park*, p. 165.

50  Eric Gable and Richard Handler, *The New History in an Old Museum (*Durham: Duke University Press, 1997), p. 85.

51  Interview with Wayne Hemingway (21 October 2013).

52  Ibid.

53  Lowenthal, 'The Past as a Theme Park', p. 12.

54  Yetter, *Williamsburg Before and After*, p. 52.

55  Frank Atkinson speech given in 1975, cited by Cross and Walton, *The Playful Crowd*, p. 223.

56  Walt Disney, cited by Bob Thomas, *Walt Disney: An American Original* (New York: Hyperion, 1994), pp. 11–13, 218. Following Disney's death, the Walt Disney Company would bring Sunset Boulevard to the masses via creation of Disney's MGM, now Hollywood Studios, in Florida.

57  Margaret J. King, 'The Theme Park: Aspects of Experience in a Four-dimensional Landscape', *Material Culture*, 34 (2002), pp. 1–4.

58  Cited by Sharon Zukin, *Landscapes of Power: From Detroit to Disney World* (Los Angeles: University of California Press, 1993), p. 222.

59  http://www.theguardian.com/artanddesign/2015/aug/20/banksy-dismaland-amuse-ments-anarchism-weston-super-mare accessed 21 August 2015.

60  Deyan Sudjic, *The Language of Things.* (London: Allen Lane, 2008), p. 9.

61  Interview with Wayne Hemingway (21 October 2013).

62  Wayne Hemingway speaking at the 'Calling All Creative Minds' event held at Turner Contemporary (April 2013).

63  James H. Gilmore and B. Joseph Pine, *Authenticity: What Consumers Really Want* (Boston: Harvard Business School Press, 2007), p. 229.

64  On this theme see Jason Wood, 'From Port to Resort: Art, Heritage and Identity in the Regeneration of Margate', in Peter Borsay and John K. Walton (eds), *Resorts and Ports: European Seaside Towns Since 1700* (Bristol: Channel View Publications, 2011), pp. 197–218.

65 Wayne Hemingway speaking at the 'Calling All Creative Minds' event held at Turner Contemporary (April 2013).
66 http://www.dreamland.co.uk/ accessed 14 June 2015; http://www.theguardian.com/travel/2015/oct/16/britains-oldest-rollercoaster-reopens-at-dreamland-margate accessed 16 October 2015.
67 Interview with Wayne Hemingway (21 October 2013).
68 Archer, 'The Limits to the Imagineered City'.
69 Wayne Hemingway speaking at the 'Calling All Creative Minds' event held at Turner Contemporary (April 2013).
70 Relph, *Place and Placelessness*, p. 65.
71 Wood, 'From Port to Resort'.

# 12  The designation of amusement parks and fairground rides in England

*Allan Brodie and Roger Bowdler*

In recent years the definition of 'heritage' has continued to evolve. No longer restricted to castles, cathedrals and country houses, the term now embraces monuments of the industrial era and institutional buildings ranging from workhouses and prisons to hospitals and schools. At the same time, there has been a growing appreciation of and public support for the heritage of the twentieth century.[1] It was in this context that English Heritage (now Historic England) chose to list the 1920 Scenic Railway at Dreamland in Margate in 2002 (upgraded to II* in 2011).[2] To date, the only other listed fairground ride is the Grade II water splash at East Park in Kingston-upon-Hull.[3]

These designations, of course, do not reflect the true wealth of the amusement park and fairground heritage that survives in England. Blackpool Pleasure Beach alone has four of the world's thirty-five surviving pre-1939 rollercoasters, while another Scenic Railway built in 1932 continues to operate at Great Yarmouth. There are other pre-1939 unlisted rides and attractions at Blackpool, including Sir Hiram Maxim's Captive Flying Machine of 1904 and the River Caves of the following year, both constructed in the sand dunes of South Shore prior to the Pleasure Beach's creation. There are also two further water splash rides at municipal parks in Kettering and Scarborough; that in Kettering being earlier than the listed example in Kingston-upon-Hull.

This chapter examines Historic England's programme of research into amusement parks and fairgrounds rides and the mechanisms for their protection that are being actively considered.[4]

## Heritage and protecting the architecture of fun

Around the turn of the twentieth century a series of measures were taken to record and protect Britain's historic environment. In 1877 the Society for the Preservation of Ancient Buildings was founded and within five years the Ancient Monuments Protection Act established the Inspectorate of Ancient Monuments.[5] The National Trust for Places of Historic Interest or Natural Beauty began work in 1895 and the Victoria County History in 1899.[6] To support the work of these bodies, the creation of national inventories of monuments was needed and so Royal Commissions for Scotland, England and Wales were established. In Scotland the date of the Act of Union in 1707 was set as the terminal date of the survey, while England's initial date was 1700, though this was extended to 1714 on the eve of the First World War. Anything dating from after 1714 was classified as 'modern' and therefore not published or recorded in detail. By 1939 the Commissioners of the Royal Commission on the Historical Monuments of England (RCHME) agreed to include a selection of eighteenth- and nineteenth-century buildings, a decision formalised after the war when a new Royal Warrant allowed them to adopt informally the terminal date of 1850.

Listing was introduced under the 1947 Town and Country Planning Act; it was a means of helping post-war reconstruction by identifying those buildings and structures that had a high claim to respect. Such claims were defined as having *special interest*. This came in two varieties: architectural and historical. From the very start, listing concerned itself with the stories and ideas around buildings as well as with their built form.[7]

A list composed only of churches, castles, country houses, cottages, civic set-pieces and the stand-out monuments of commerce and industry would be earnest, deserving and sensible, but it would miss out on much of the spice of life, and be glaringly incomplete. The notion that pleasure and fun are fitting topics for study is now well established; Dutch historian Johan Huizinga's ground-breaking study of the history of play in society, *Homo Ludens*, first appeared in English in 1949, only two years after the 1947 Town and Country Planning Act.[8] The listing of cinemas and pubs is now normal, but it was not always so. Cinemas began to be listed in increasing numbers during the urban re-surveys of the 1980s. English Heritage began a thematic survey of pubs, ushered in by the publication of its booklet *Pubs: Understanding Listing* (1994). In recent years, English Heritage's, and since 2015 Historic England's, listing programme has further celebrated and embraced popular culture. The designation in 2013 of Manze's pie and mash shop in Walthamstow was not the first excursion into the protection of working-class eateries, but it was still remarkable enough to warrant a ministerial visit.[9] Beatles-graced zebra crossings, Beatnik basements, model villages, diving boards and skateboard parks have all made it onto the National Heritage List for England (NHLE).[10] Making sure that the NHLE reflects all facets of life is one of guiding principles of today's approach to designation.

Seaside resorts are a case in point. Although 10 per cent of all designated assets are reckoned to lie within a mile of the coast, there is always more to do if the designation base is to be a true reflection of our heritage, ancient or modern. Work has now taken place at a number of seaside resorts to begin to bring the coverage of the NHLE more up to date. In 2007 a research project on seaside resorts culminated with books on the evolution of the seaside resort and the seaside heritage of Margate, followed in 2008 and 2014 by books on Weymouth and Blackpool.[11]

One clear finding from the work undertaken in recent years by English Heritage on seaside resorts is that while the lists, most of which date from the 1980s, are very good at capturing pre-1840 heritage, they are rather less good at incorporating more recent buildings, particularly those of the leisure industries. This is a consequence of the legal framework within which listing has to take place.[12] Older buildings are inevitably rarer and therefore more likely to be of special interest. Therefore, before 1700, all buildings containing a significant proportion of their original fabric are listed; from 1700 to 1840, most buildings are listed; but after 1840, because so many more buildings were erected and have survived, greater selection is required, especially for buildings dating from after 1945. Most of England's seaside heritage post-dates 1840 and therefore it is necessary to determine its originality, its aesthetic quality and its special interest.

Approaches to the selection of buildings and structures for listing are described in English Heritage's *Designation Selection Guide: Culture and Entertainment* (2011).[13] Where internal disputes have arisen over the listing of such structures, they tend to centre on perceptions of authenticity. When does a much-replaced apparatus cease to be historic? Seaside structures are particularly prone to change. The damp and salty environment attacks fabric, while commercial considerations, health and safety, wear and tear all lead to further alteration. At what point do these considerations fatally undermine authenticity? Heritage is distinctly suspicious of the replica.

Another issue is whether the architecture of fun should be brought into the planning system anyway? Is not adaptability and updating at the heart of commercial success? Are not these structures simply too ephemeral to be to be relevant? This is where the *historical* aspect of designation comes to the fore and where the fundamental point of listing comes into its own. Listing is about much more than triggering an extra degree of oversight in the management of change, through the planning system, vital though that is. It is a celebration of special interest, a flag of recognition that some buildings and structures (or 'heritage assets' in official parlance[14]) warrant recognition and deserve our attention.

English Heritage's *Conservation Principles* can help here too.[15] The fundamental values that contribute to overall significance are *evidential, historical, aesthetic* and *communal*. Consideration of the nature of significance should inform our response to applications for change; establishing the grounds of importance will make clearer the route to acceptable proposals. The aesthetics of a centuries-old brick wall would demand a rather different conservation response to the treatment of a painted concrete Modern Movement wall. Remembering the *historical* significance of a structure may enable us to see beyond the *aesthetic* detractions of an altered structure, and remember the *communal* value of a once-popular leisure structure, and recognise signs of change as of note in offering *evidential* insight into its evolution. This approach serves as a helpful corrective to the tendency towards judging candidate assets purely on visual grounds. It serves as a reminder of the importance of associations, alongside the fabric, which reflects the original guiding principles of listing in the 1940s, when historical interest was established as one of the twin grounds for listing.

## Research on rollercoasters and amusement parks

In 2002 the Scenic Railway at Dreamland in Margate was listed Grade II (Figure 12.1). The ride was the centrepiece of the amusement park that opened in

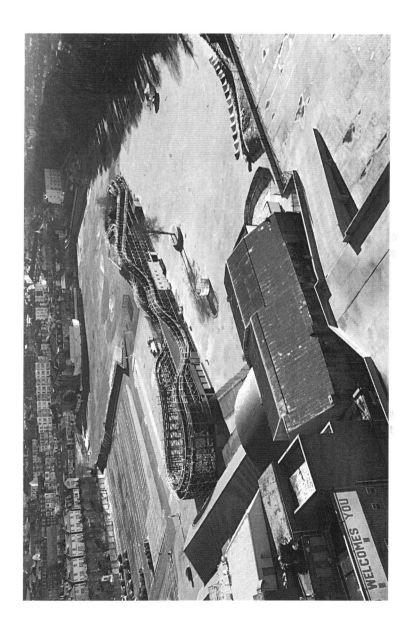

*Figure 12.1* The Scenic Railway at Dreamland, Margate in 2007, a year before the fire. The rest of the park had been cleared by this date

Source: Peter Williams © Historic England (DP032138)

1920 on the site of the Hall-by-the-Sea entertainment complex and former menagerie of the self-ennobled 'Lord' George Sanger's pleasure gardens.[16] The Scenic Railway is a wooden, side-friction rollercoaster with cars on which a brakeman rides to control its speed at fast corners. The track is approximately three-quarters of a mile long and each train of three cars could take up to twenty-eight passengers, plus the brakeman. The cars run on rails set on to a wooden bed supported by major horizontal timbers. They are guided along the track by side-friction wheels running against boards along the side of the track. The major elements are bolted together, but smaller struts and braces are nailed to the major timbers.

In 2008, the engine shed, where the trains were stored and repaired, was destroyed in an arson attack, but the motor house, its two electric motors, large manual clutch and the belt drive mechanism for conveying the power to the chain on the lift hill all survived. Following the fire there was a need to reassess the Scenic Railway's listing. This involved re-examining briefly the survival of historic rides in Britain and abroad to assess its significance, while at the same time initiating research to understand the evolution of amusement parks. The result of this research was two-fold. First, in 2011, despite the damage, the decision was taken to upgrade the Scenic Railway to II*, making it one of the 5.5 per cent of historic buildings accorded this grading.[17] Second, was the recognition that England had a particularly rich heritage of amusements and rollercoasters, which was inadequately appreciated and protected. In the emerging English Heritage Action Plan for the National Heritage Protection Plan for 2011–15 resources were allocated to allow a national review of Amusement Parks and Fairground Rides and this work took place in 2013.[18]

The project involved assessing what was already designated and what should be considered for future designation. Therefore the key questions were:

- What existed?
- What remains and how complete is it?
- How rare are the surviving remains and how does England's remaining amusement park heritage compare with what remains internationally?

## What existed and what remains

The first task was to understand the origins and evolution of the amusement park in England. The surviving seaside parks in Blackpool and Great Yarmouth, and at Dreamland in Margate, were founded in the late nineteenth and early twentieth centuries, but the origins of the modern amusement park can be traced back to a number of British antecedents as well as the first amusement parks across the Atlantic.

The first strand in their story in Britain is the annual fairs that were held in towns and villages where goods were traded, though increasingly they also provided opportunities for England's rural population to gather and enjoy entertainment (Figure 12.2). One author described the evolution of fairs as: 'their origin was religious, their development commercial, and their apotheosis an unrestrained indulgence in pleasure or license, as you may choose to regard their diversions'.[19]

*Figure 12.2* Witney Fair, Henry W. Taunt, 1904

Source: © Historic England (CC73/00569)

Although fairs were a temporary if regular annual event, there is evidence that some permanent structures were created. Stourbridge Fair (Worcestershire) was a notable commercial market and an occasion for:

> Coffee-Houses, Taverns, Eating-Houses, Music Shops, Buildings for the Exhibition of Drolls, Puppet Shews, Legerdeman, Mountebanks, Wild Beast, Monsters, Giants, Rope Dancers, etc . . . Beside the Booths, there are six or seven brick Houses . . . and in any of which the Country People are accommodated with hot and cold Goose, roast or boiled Pork, etc.[20]

At Weyhill (Hampshire) a range of former hop growers' fair booths, which date from the first half of the nineteenth century, enclose an elongated U-shaped yard.[21] They were erected by a consortium of hop growers to market their produce and are the last vestiges of Weyhill Fair, first documented in 1225.[22]

Circuses and menageries became a regular feature of fairs by the early nineteenth century. In the second half of the nineteenth century the aforementioned 'Lord' George Sanger became the most prominent circus proprietor and menagerie owner and through his establishment of a permanent base in Margate has a direct role in the origins of Dreamland and the modern amusement park.[23]

Since the seventeenth century small rides powered by men or horses had appeared at some fairs.[24] The logical technological step was to apply steam power to these rides and by the 1860s firms involved in the production of farm machinery were beginning to do this.[25] Subsequently, the internal combustion engine and electricity further transformed fairgrounds, allowing larger, faster and more brilliantly illuminated rides. This new technology required significant investment and therefore showmen increasingly concentrated on larger fairs, resulting in the decline of the smaller, traditional, village fair.[26] It would also lead to the creation of the earliest, purpose-built, permanent amusement sites.

A second indigenous strand in the origins of modern seaside amusement parks is the pleasure garden, a phenomenon that can ultimately trace its roots back to the pleasure gardens of medieval monarchs.[27] The earliest public pleasure garden had been created in London during the seventeenth century. New Spring Gardens, later called Vauxhall Gardens, opened soon after the Restoration and by the eighteenth century London was said to be home to more than sixty.[28] Pleasure gardens offered their paying customers musical performances, fireworks and at the end of the eighteenth century even balloon ascents, the ultimate novelty ride.[29] The pleasure gardens of London were the inspiration for similar institutions in major provincial towns and at seaside resorts. One of the earliest was at Margate where the remains of the medieval Dent-de-Lion fortified house at nearby Garlinge became a destination for visitors by the late eighteenth century. Margate's Tivoli Gardens opened in 1829 providing a spacious park with a concert hall, a bowling green, an archery ground, refreshment rooms, arbours, Swiss cottages and sequestered walks.[30] At Broadstairs the Ranelagh Gardens at St Peter's was designed as a replica of the Tivoli Gardens as well as seeking to emulate its metropolitan namesake.[31]

In the 1860s Blackpool was beginning to provide its visitors with pleasure grounds. Belle Vue Gardens was created as an inland destination for visitors and in the early 1870s a larger, more modern rival opened nearby. Raikes Hall Gardens opened for its first full season in 1872 and provided its customers with firework displays, circus acts, dancing, acrobats and a range of other lively spectacles.[32] The gardens contained a substantial conservatory filled with exotic plants, ferns and flowers, as well as a Grand Pavilion capable of accommodating 10,000 people and a dancing platform that catered for 4,000 people.[33] During the next twenty-five years a series of new attractions were added including a tricycle track, a camera obscura and a switchback railway.[34] Raikes Hall Gardens was in effect a proto-amusement park, a few miles from where Blackpool Pleasure Beach would develop in the next few years.

The amusement park was also inspired by early national and international exhibitions and many early switchbacks, scenic railways and figures of eight were originally located at short-lived exhibition sites before being transferred to amusement parks. In London, the most important exhibition was the Great Exhibition in 1851, which did not feature amusements, but exhibitions at Earl's Court, Kensington's Olympia and Shepherd's Bush's White City all had areas containing amusements.[35]

While some of the strands behind the origins and the popularity of amusement parks is indigenous to Britain, the most immediate and direct debt was to developments taking place across the Atlantic at Coney Island in Brooklyn (New York), where the earliest amusement parks were established from the mid-1890s onwards. In 1895 Captain Paul Boyton opened Sea Lion Park, the first enclosed amusement park with an admission fee.[36] By the early twentieth century three vast parks lit by hundreds of thousands light bulbs had been built along the seafront; Steeplechase (opened in 1897), Luna Park built on the site of Boyton's first, unsuccessful park (1903) and Dreamland (1904).[37] Today, Coney Island no longer glitters so brightly, but the footprints of the parks are still evident. The Cyclone, a wooden rollercoaster of 1927, still operates and is rated by enthusiasts as one of the greatest rides in the world. It is also a protected monument, one of a handful of fairground rides to have been afforded this recognition in America.[38]

Having examined the origins of amusement parks, the research examined how Blackpool Pleasure Beach served as a blueprint for amusement parks in Britain. The site of Britain's first enclosed seaside amusement park evolved from the mid-1890s onwards on a stretch of the shoreline to the south of the Victoria Pier, an area occupied by a gypsy encampment. The first attractions in this area were small-scale rides including a Hotchkiss Railroad Bicycle Ride, as well as fairground booths and stalls.[39] These were created as concessions on plots of land in the dunes and the title of 'The Pleasure Beach' only began to appear in advertisements in 1905, indicating that the process of consolidating the ownership of the plots and rides was under way.[40]

Blackpool Pleasure Beach served as an inspiration and a model for other seaside amusement parks, like the Kursaal and Pleasureland. The Kursaal at

Southend-on-Sea opened in July 1901, with gardens, amusements, a cycle track, a café and a menagerie and circus.[41] The site changed ownership in 1910 and became Luna Park with the Harton Scenic Railway and a figure of eight coaster, a miniature railway, Astley's Circus and a cinema.[42] The public face of the Kursaal was a large, red-brick structure with ashlar detailing and dominated by a tall, 'Wrennaissance'-style dome. There was a plan to erect a 530-foot tower, but this was never built. Pleasureland at Southport has been operating since 1913, when it opened with a figure of eight rollercoaster and a slide.[43] The park, which closed in 2006, included several historical rides, including the Cyclone. Erected in 1937, this structure had undergone some previous losses, but was substantially intact in September 2007, when requests for listing reached English Heritage. By the time an official site visit took place, the structure lay in pieces (Figure 12.3): 'one of the most shocking closures in recent years'.[44]

Apart from Blackpool Pleasure Beach, the most famous seaside amusement park is Dreamland at Margate. At the end of 1919 the site was sold to John Henry Iles who was marketing rollercoasters for L. A. Thompson's Scenic Railway Company in Britain and Europe.[45] It opened again in April 1920 with an amusement park and cinema and was called Dreamland, to evoke some of the glamour of Coney Island.[46] During the 1920s Charles Palmer, a local architect, improved the park's appearance by replacing lightweight, wooden structures with a unified scheme in brick and concrete.[47] In 1935 the present Dreamland Cinema was opened and contained a 2,200 seat auditorium and a multi-entertainment complex.[48] First listed in 1992, it was upgraded to Grade II* in 2007, despite having undergone some losses and changes, such as sub-division of the auditorium. This was in recognition of its architectural quality, the importance of the sculpted decoration by Eric Aumonier, and its overall contribution to the Dreamland complex. Dreamland closed during 2006 after decades of changing ownership, under-investment and declining visitor numbers. By this time all the fixed rides, with the exception of the Scenic Railway, had been removed.

Seaside amusement parks have been in a serious decline in England, with Blackpool, Great Yarmouth and Southport (re-opened) being the only traditional parks that have survived. There are also some modern parks, such as the large Fantasy Island at Ingoldmells and the Brean Leisure Park near Weston-super-Mare. In recent years there has also been a trend towards providing smaller parks nearer the heart of resorts such as at Scarborough beside the harbour or near the pier at Southend-at-Sea. In large towns and cities urban amusement parks have also been in decline. Instead, today the majority of the larger amusement park sites are at inland parks, on country house estates that sought to diversify to increase their income. These include Alton Towers, whose elaborate gardens opened to the public in 1860.[49] During the Second World War it was requisitioned by the military and remained closed to the public until 1951 when fairground attractions were first provided. From 1973 the estate was run as an amusement park and in 2010 attracted 2.6 million visitors.[50] Drayton Manor Park at Tamworth, which was Sir Robert Peel's home, became Drayton Manor Park and Zoo in the 1970s and in the 1980s became a theme park.[51] Chessington World of Adventures had

*Figure 12.3* The Cyclone at Pleasureland in Southport, 2005

Source: © Historic England (DP034506)

different origins; it began as a zoo in 1931 and became one of Britain's first theme parks in the mid-1980s.[52]

Fairground rides were not restricted to amusement parks during the nineteenth and twentieth centuries and there were many examples of single rides set up on seafronts or on any convenient piece of land. A story in an 1823 magazine includes an early reference at Margate to a slide called the Russian Mountains near the harbour and an advertisement in October 1823 announced that a Montagne Russe at Sadler's Wells would soon be closing.[53] At Blackpool the 1893 Ordnance Survey map shows a switchback near the Central Pier and another existed to the north of Blackpool at Uncle Tom's Cabin. At the Winter Gardens a large Ferris wheel opened in 1896 and this remained an iconic feature of Blackpool until it was dismantled in the late 1920s.[54] At Folkestone a switchback was erected on the beach in 1888, but succumbed in 1919–20 to a major fire.[55] Piers might also contain rides, such as the switchback filling most of the deck of Ramsgate Pier, which operated from 1888 until 1891.[56] Birnbeck Pier at Weston-super-Mare offered its customers swings in 1876 and by the early twentieth century it had a switchback, a water chute, a helter skelter and a short-lived flying machine. Constructing so many entertainment facilities was possible as the pier included an island in its structure.[57]

## Significant survivals and later developments

Having reviewed the background to the origins of amusement parks, and assessed the form of the modern amusement park market, the assessment examined what still survives today that might merit heritage protection measures by Historic England.

The most important amusement park site in England is Blackpool Pleasure Beach. With its unique collection of early rides, nowhere else in the world so clearly reflects the past 120 years of amusement parks and the evolution of rollercoaster and fairground ride technology. It has an unrivalled heritage of pre-1939 rides stretching from Sir Hiram Maxim's Captive Flying Machine (1904) and the River Caves (1905) to the Big Dipper (1923), the Roller Coaster (now the Nickelodeon Streak, 1933), the Little Dipper (now the Blue Flyer 'kiddie coaster', 1934) and the Grand National (1935), Britain's only surviving historic twin-track rollercoaster.[58]

Sir Hiram Maxim's Captive Flying Machine, the oldest ride in continuous use in the world, first operated in Blackpool on 1 August 1904 and therefore predates the foundation of the Pleasure Beach.[59] Devised in 1902, it was first shown at the Earl's Court exhibition in 1903 and was rebuilt at Blackpool.[60] It consists of ten steel arms from which cables hang to support cars, and these rotate around a central 30m-high vertical driving shaft, allowing the cars to fan outwards as they turn, achieving what was once a terrifying maximum speed of 65 mph.[61] The original gondolas were replaced in 1929 by aeroplanes and rockets replaced these in 1952, updates designed to keep the ride at the forefront of contemporary technology. In 1905 the River Caves opened.[62] The ride became popular in 1904 in America and had come to England first at Earl's Court. The ride consisted of boats passing through 'caverns' with tableaux lit by electric lights. The ride has survived though the interiors have been updated. Other pre-1914 rides have not alas survived. In 1907 John Henry Iles, who later founded Dreamland, erected the Scenic Railway, which he continued to own until his bankruptcy in 1919.[63] The 65 foot-high Water Chute was also erected in 1907, propelling fifty-five boats per hour, each with their own gondolier, down a 267 foot-long chute.[64] In 1909 the Velvet Coaster was built by William H. Strickler. It remained in use until 1932 when it was dismantled, though elements of it were incorporated into the Roller Coaster of 1933.[65]

The programme of providing new rides for the Pleasure Beach continued during the inter-war years and a few have survived. Noah's Ark, also constructed by Strickler, invited visitors to take a trip to Mount Ararat, walking past animals on moving platforms with a rocking motion to simulate the sea voyage.[66] It survives today, with renewed animals, and is a key feature near the main entrance to the park. The Big Dipper, the first example of which appeared in New Jersey in 1920, was designed by the leading rollercoaster engineer John A. Miller, and was added to the Pleasure Beach in 1923, again by Strickler.[67] By the mid-1930s extra land was available to the south of the Pleasure Beach and so the Big Dipper was lengthened and rearranged by the American engineer Charles Paige in 1936, while the new station was created by the British architect Joseph Emberton. In 1933 Paige oversaw the construction of the Roller Coaster (now the vibrant-orange Nickelodeon Streak), reusing the lift hill and other parts of the aforementioned Velvet Coaster.[68] In 1934 the Little Dipper (later known as the Zipper Dipper, now the Blue Flyer) was built and a year later Paige working with Harry G. Traver built the Grand National, a ride inspired by Travers' Cyclone at Long Beach (California).[69] In total, Paige is known to have been responsible for thirteen wooden rollercoasters, but the rides at Blackpool are his only creations that are still in operation.[70]

As well as importing rides and ride technology, the idea of a visual theme for the park seems to have been influenced by American parks. In 1931 the Philadelphia architect Edward Schoeppe designed a new frontage to the park, including the 600-seat News Theatre, as well as the front of the Velvet Coaster and a new frontage for Howell's Photographic Studio, which he wittily designed to resemble a camera.[71] Beginning in 1933, and continuing until the outbreak of the Second World War, Emberton was employed to give the park an overall visual unity, which involved creating more modernist structures and gave existing rides a Moderne facelift (Figure 12.4).[72]

Since the Second World War Blackpool Pleasure Beach has updated its offer to its customers with new rides and attractions being squeezed in alongside many favourite historic rides to create an increasingly complex entertainment landscape. The major post-war steps in the creation of this modern wonderland began with the creation of the last wooden rollercoaster, the Wild Mouse, in 1958 and the fast carousel, the Derby Racer in the following year.[73] They were followed in 1967 by the Log Flume, which was 50 feet high and almost 2,350 feet long making it the longest outside the USA.[74] The Steeplechase was opened in 1977 and two years later the Revolution was created, a steel shuttle rollercoaster that provides

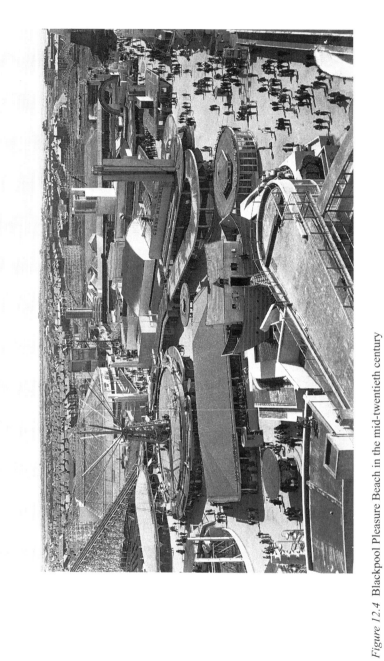

*Figure 12.4* Blackpool Pleasure Beach in the mid-twentieth century

Source: © Historic England Aerofilms Collection (AFL03/Lilywhites/blp45)

a looping ride on its outward and return journey. The Avalanche, which opened in 1988, is a steel bobsled rollercoaster, the first of its kind in Britain, but all these were literally overshadowed by the 1994 Pepsi Max Big One, now the Big One. It was designed by Ron Toomer of Arrow Dynamics and is 235 feet high, one mile long and reaches a maximum speed of 74 mph.[75] In 2000, Valhalla, the world's largest indoor ride, opened combining a white knuckle ride with special effects creating the appearance of fire, water, snow, thunder and lightning.[76] What is remarkable about Blackpool Pleasure Beach is the ingenuity of its designers and the foresight of its management to realise new attractions while retaining the best of the park's historic rides.

In addition to Blackpool and Margate, another pre-Second World War rollercoaster survives at Great Yarmouth. The first amusement ride, a L. A. Thompson Switchback Railway, was erected in Great Yarmouth on the seafront in 1887 and remained there until the early twentieth century.[77] In 1909 the Pleasure Beach was established and included a Scenic Railway set within a plaster mountainous terrain, though it was destroyed by fire in 1919 but rebuilt quickly. An Aerofilms photograph of 1920 shows the Scenic devoid of any scenery, standing alone on the sand.[78] In 1928 the lease of the Scenic Railway ended and the ride was transferred to Aberdeen. A new Scenic Railway was purchased from Paris and opened in 1932 with an alpine landscape and castles entertaining riders on the circuit. The ride was re-clad in the 1960s.

As well as historic rides surviving in amusement parks a number of towns have parks that include small rides such as miniature railways and water chutes. Charles Wicksteed, the founder of Wicksteed Park at Kettering, created a 30-acre lake in 1921 and later designed and built one of the earliest surviving water-based rides in the world, the Water Chute, which opened in 1926.[79] It plunges the occupants of a boat down a ramp into the water at speed. The example in East Park at Kingston-upon-Hull was designed and made in 1929 by Messrs Charles Wicksteed and Co. Ltd and erected by the City Council's Engineer's Department.[80] A third example survives in Peasholme Park in Scarborough. Only the ride at Hull is listed, though all three are in a similar state of survival and of a similar date.

The tradition of single rides in prominent locations continues at seaside resorts. On the Promenade at Southport, at the landward end of the pier, a set of restored gallopers has become a semi-permanent feature of the resort's seafront, and a number of resorts, as well as inland towns and cities, have been the host for large Ferris or observation wheels, a phenomenon stimulated by the success of the London Eye.

Today, some seaside piers are also the home to fairground rides. With limited space available these are often aimed at younger riders, but Blackpool's South Pier offers the Sky Coaster and the Sky Screamer, rides that catapult the bravest high into the sky.

## International comparison

Although Britain has a rich amusement heritage, this is only a tiny fraction of the dozens of rides that once existed. Trawling through various sources, it is possible make rough estimates of the original numbers, though there are undoubtedly some that may have been missed.[81] Nevertheless, there seem to have been thirty-one switchback railways, though three of these were dismantled and relocated to new locations. None survive today anywhere in the world and the last British example was demolished in the 1930s. In Britain there were thirty-eight side-friction rollercoasters and thirty-four figure of eight rides. While none of these survive, the side-friction technology used in Scenic Railways can still be enjoyed at Great Yarmouth and now again at Margate; the remaining two of Britain's original thirty-one examples.

Britain's earliest rollercoaster is the Scenic Railway at Margate dating from 1920, meaning that it is the sixth equal oldest surviving ride in the world, along with the Jack Rabbit at Kennywood, West Mifflin (Pennsylvania). The oldest ride still in use is the 1902 side-friction Leap the Dips at Lakemont Park, Altoona (Pennsylvania) while the oldest scenic railway, and the oldest rollercoaster which has been in use continuously is at Luna Park in Melbourne, Australia, dating from 1912.[82] Europe's oldest rollercoaster ride is the Rutschebanen at Tivoli Gardens in Copenhagen, which still has mountain scenery surrounding the coaster. In total, thirty-five rollercoasters have survived from before 1939, twenty-three in the USA and six in Britain, including four at Blackpool Pleasure Beach.

Four rides in the USA enjoy statutory protection by being on the National Register of Historic Places (NRHP). In California, the 1925 Mission Beach Giant Dipper rollercoaster at Belmont Park San Diego, and the 1911 Looff Carousel and 1924 Giant Dipper rollercoaster on the Santa Cruz Beach Boardwalk, are also designated.[83] The most famous and most celebrated rollercoaster is the 1927 Cyclone rollercoaster at Coney Island, which is claimed by enthusiasts to be the greatest ride in the world.[84]

## The challenge of retention

Historic amusement park rides are rare survivors and face significant threats. Potentially most destructive is the risk of fire damage, as these often fragile structures were usually constructed of wood. An 1888 switchback on the beach at Folkestone succumbed spectacularly to a major fire in 1919 or 1920 and Dreamland at Margate has suffered a number of fires during its history, most recently the arson attack on the Scenic in 2008.[85] In addition, all fairground rides will have been subject to programmes of maintenance and repair to guarantee their survival. This is especially the case with timber structures and this has inevitably led to the loss of original, historic fabric. Despite comprehensive and regular maintenance programmes, they must inevitably have a limited lifespan as rides (averaging 20–40 years) because over time their structure becomes beyond economic repair and ultimately unsafe as their fabric ages and regulations become stricter. In addition, some have had their appearance updated to respond to changes in public tastes. As mentioned earlier, the customers flying on Sir Hiram Maxim's Captive Flying Machine in 1904 were transported in gondolas, which were replaced in 1929 by aeroplanes and in turn by rockets in 1952.

The technology of rides developed rapidly during the twentieth century rendering many older rides old fashioned and potentially less popular. However, age also confers a certain nostalgic value to rides and those once aimed at thrill-seeking adults have been successfully re-imagined as fun rides for the family. Blackpool Pleasure Beach's rich heritage of rollercoasters and other early attractions demonstrate that this policy can be a success.

Although they are well loved by the public and are recognised by the management of the Pleasure Beach as a key part of what makes the Pleasure Beach special, none of the rides at Blackpool are as yet listed structures, though their designation is actively under consideration at the time of writing (November 2015). As long as the Pleasure Beach remains popular and economically successful, the future of these rides should be secure, but if there is a belief that a new ride would increase visitor numbers, even at the cost of an historic ride, these could be lost without any consultation with the public, local authority or Historic England. Listing is a mandatory duty: if special architectural or historic interest is present, then the Secretary of State has a duty, not the option, to add the building to the list. Thankfully, designation is just the start of a conversation about the management of an asset, and not the petrifying closure of discussion which misinformed opinion sometimes mistakes it for.

## Conclusion

This chapter has sought to demonstrate that there is a place for listing the architecture of fun. If an understanding of amusement park and fairground heritage can be a major contributor to the renaissance of the English seaside, then there is a need to identify the components of this legacy, and, where appropriate, deliver protection through designation. It is unlikely that many more such structures will be added to the NHLE, but there is a need to ensure that the most important survivors gain the recognition that listing bestows: designation is a celebration of special interest, as well as a tool in the planning system for careful management. And with governmental encouragement of the identification of local heritage assets, there is no reason why other levels of protection cannot be bestowed as well. Designation in itself will not make holidaymakers line up in droves, though there are many people who would prefer historic rides to modern thrill rides. However, a celebration of the special, and imaginative management of this select group, will guarantee that the best of our seaside pleasure inheritance gets every chance to see many more holiday seasons yet.

## Notes

1 For an online, searchable database of official records of all nationally designated heritage assets, see the National Heritage List for England (hereafter cited as NHLE): https://historicengland.org.uk/listing/the-list/.
2 See Nick Laister, this volume.
3 The Aerial Glide Static Fairground Ride at Shipley Glen Amusement Park in Baildon was delisted when it was discovered that its steel frame was a later replacement.

4  For a guidance note on the ongoing research see Allan Brodie, *Historic Amusement Parks and Fairground Rides: Introductions to Heritage Assets* (London: Historic England, 2015): http://historicengland.org.uk/images-books/publications/iha-historic-amusement-parks-fairground-rides/ accessed 15 September 2015.

5  Andrew Sargent, 'RCHME 1908–1998: A History of the Royal Commission on the Historical Monuments of England', *Transactions of the Ancient Monuments Society* 45 (2001), pp. 57– 80.

6  Established to preserve and celebrate the rural heritage of England, which was under pressure from industrialisation, the VCH's metropolitan counterpart was the Survey of the Memorials of Greater London (later the Survey of London).

7  Simon Thurley, *Men from the Ministry. How Britain Saved its Heritage* (New Haven and London: Yale, 2013), pp. 200ff.

8  Johan Huizinga, *Homo Ludens: A Study of the Play-element in Culture* (London: Routledge & Kegan Paul, 1949).

9  L. Manze's eel, pie and mash shop at 76, Walthamstow High Street, London Borough of Waltham Forest was listed on 30 October 2013 (NHLE 1416834).

10 The Abbey Road crossing, renowned as the cover of the Beatles' 1969 album *Abbey Road* was listed in 2010 (NHLE 1396390). The Casbah Club at 8, Hayman's Green, Liverpool 12, was owned by Pete Best's mother, and decorated by Best, John Lennon and friends; it ran from 1959 to 1962, and was listed in 2006 (NHLE 1391759). The Model Village at the Old New Inn, Bourton-on-the-Water, Gloucestershire, was constructed in 1936–40 and listed in 2013 (NHLE 1413021). The Coate diving platform at Coate Water, Swindon, was built in 1935 and listed in 2013 (NHLE 1417099). The Rom Skate Park, on Upper Rainham Road, Hornchurch, London Borough of Havering, was designed by skateboarder Tony Rolt along the lines of Californian exemplars and opened in 1978. It was listed in August 2014 (NHLE 1419328).

11 Allan Brodie and Gary Winter, *England's Seaside Resorts* (London: English Heritage, 2007); Nigel Barker, Allan Brodie, Nick Dermott, Lucy Jessop and Gary Winter, *Margate's Seaside Heritage* (London: English Heritage, 2007); Allan Brodie, Colin Ellis, Stuart David and Gary Winter, *Weymouth's Seaside Heritage* (London: English Heritage, 2008); Allan Brodie and Matthew Whitfield, *Blackpool's Seaside Heritage* (London: English Heritage, 2014).

12 https://www.gov.uk/government/uploads/system/uploads/attachment_data/file/137695/Principles_Selection_Listing_1_.pdf accessed 13 August 2014.

13 https://historicengland.org.uk/images-books/publications/dlsg-culture-entertainment/ accessed 13 August 2014.

14 This term has been used since the publication by the Department for Communities and Local Government of the *National Planning Policy Framework* in 2012.

15 *Conservation Principles. Policies and Guidance for the Sustainable Management of the Historic Environment* (London: English Heritage, 2008): https://www.english-heritage.org.uk/professional/advice/conservation-principles/ accessed 14 October 2014.

16 See Nick Laister, this volume. Also R. V. J. Butt, *The Directory of Railway Stations* (Sparkford: PSL Ltd, 1995), p. 155; Richard Clements, *Margate in Old Photographs* (Stroud: Alan Sutton, 1992), p. 125; Nick Evans, *Dreamland Remembered* (Whitstable: Bygone Publishing, 2009), pp. 5–6.

17 NHLE 1359602. A useful comparison here can be made with the regeneration of seaside piers, which repeatedly undergo replacement of components. Clevedon Pier is a classic example; first listed in 1971, it was upgraded in 2001 to Grade I after the completion of a major restoration programme funded by the Heritage Lottery Fund: see NHLE 1129687.

18 https://www.english-heritage.org.uk/professional/protection/national-heritage-protection-plan/ accessed 27 August 2014.

19 William B. Boulton, *Amusements of Old London*, 2 vols (London: Frederick Muller Ltd, 1971), pp. 2, 44.

20  Robert W. Williamson, *Popular Recreations in English Society 1700–1850* (Cambridge: Cambridge University Press, 1973), pp. 20–21.

21  English Heritage Archive Building File 96389.

22  NHLE 1253842. Thomas Hardy immortalised Weyhill Fair as Weydon Priors Fair in *The Mayor of Casterbridge* at which Michael Henchard is said to have sold his wife: Thomas Hardy, *The Mayor of Casterbridge* (Oxford: Oxford University Press, 2008), p. 105.

23  Brenda Assael, 'Sanger, George (1825?–1911)', *Oxford Dictionary of National Biography* (Oxford: Oxford University Press, 2004); online edn, January 2008: http://www.oxforddnb.com/view/article/35940 accessed 14 August 2014. Remarkably, a number of the lion cages, built by Sanger after 1874, survive on the west perimeter wall of Dreamland and were listed in 2009 (NHLE 1392931). They are a poignant reminder of the cramped conditions that animals were subjected to in Victorian times.

24  Ian Starsmore, *English Fairs* (London: Thames and Hudson, 1975), p. 16.

25  Ibid., pp. 46–9; Hugh Cunningham, *Leisure in the Industrial Revolution* (New York: St Martin's Press, 1980), p. 174; Peter Wilkes, *Fairground Heritage* (Burton-upon-Trent: Trent Valley Publications, 1989), pp. 42, 47.

26  Cunningham, *Leisure in the Industrial Revolution*, p. 174; Wilkes, *Fairground Heritage*, p. 47.

27  John Harvey, *Medieval Gardens* (London: B. T. Batsford & Co., 1981), p. 106; Elaine Jamieson and Rebecca Lane, 'Monuments, Mobility and Medieval Perceptions of Designed Landscapes: The Pleasance, Kenilworth', *Medieval Archaeology* 59 (2015), pp. 255–71.

28  T. J. Edelstein (ed.), *Vauxhall Gardens: A Catalogue* (New Haven: Yale, 1983), p. 11; Jeremy Black, *A Subject for Taste* (London: Hambledon, 2005), p. 198; Mollie Sands, *The Eighteenth-Century Pleasure Gardens of Marylebone* (London: Society for Theatre Research, 1987), p. 11; Sarah Jane Downing, *The English Pleasure Garden 1660–1860* (Oxford: Shire Publications, 2009), p. 11; David Coke and Alan Borg, *Vauxhall Gardens: A History* (New Haven: Yale, 2011), p. 18.

29  Cunningham, *Leisure in the Industrial Revolution*, p. 95. Charles Dickens witnessed balloon ascents in Vauxhall Gardens: Charles Dickens, *Sketches by Boz* (London: MacMillan, 1958), pp. 120–21. A balloon ascent still takes place near the seafront at Bournemouth: http://www.bournemouthballoon.com/ accessed 14 August 2014.

30  Malcolm Morley, *Margate and its Theatres, 1730–1965* (London: Museum Press, 1966), p. 52; G. E. Clarke, *Historic Margate* (Margate: Margate Public Library, 1975), p. 51.

31  Morley, *Margate and its Theatres*, p. 55. The Ranelagh Gardens included a chance to see imported natives from Patagonia in their natural habitat! Unusually, some of the buildings have survived and are listed (NHLE 1222837, 1222976, 1223056, 1267679).

32  John K. Walton, *Blackpool* (Edinburgh: Edinburgh University Press, 1998), p. 88.

33  *Blackpool Gazette and News* (30 May 1873), p. 1.

34  The 1893 Ordnance Survey map shows the substantial facilities available at the renamed Royal Palace Gardens. The much smaller Belle Vue Gardens appear a short distance to the east.

35  See Josephine Kane, this volume.

36  Robert Cartmell, *The Incredible Scream Machine* (Ohio: Amusement Park Books, 1987), p. 66; Mark Wyatt, *White Knuckle Ride* (London: Salamander, 1996), p. 13; Gary S. Cross and John K. Walton, *The Playful Crowd: Pleasure Places in the Twentieth Century* (New York: Columbia University Press, 2005), p. 39.

37  David Bennett, *Roller Coaster* (London: Aurum, 1998), pp. 19–21; Cartmell, *The Incredible Scream Machine*, p. 66; Cross and Walton, *The Playful Crowd*, pp. 39–42.

38  National Register Historic Places (hereafter cited as NRHP) 91000907.

39  John K. Walton, *Riding on Rainbows: Blackpool Pleasure Beach and its Place in British Popular Culture* (St Albans: Skelter Publishing, 2007), pp. 21–2.

40  Bennett, *Roller Coaster*, p. 23; Robert E. Preedy, *Roller Coasters: Their Amazing History* (Leeds: Robert E. Preedy, 1992), p. 8; Walton, *Riding on Rainbows*, pp. 16ff; Josephine Kane, 'A Whirl of Wonder!' British Amusement Parks and the Architecture of Pleasure 1900–1939' (unpublished PhD thesis, University of London, 2007), pp. 74–5; Vanessa Toulmin, *Blackpool Pleasure Beach: More Than Just an Amusement Park* (Hathersage: Boco, 2011), pp. 11, 15; Josephine Kane, *The Architecture of Pleasure: British Amusement Parks 1900–1939* (Farnham: Ashgate, 2013), pp. 31–2.

41  Ken Crowe, *Kursaal Memories: A History of Southend's Amusement Park* (St Albans: Skelter Publishing, 2003), p. 7.

42  Kane, 'A Whirl of Wonder!', pp. 108–10; Kane, *The Architecture of Pleasure*, p. 63.

43  http://www.joylandbooks.com/themagiceye/articles/goinggoinggone.htm accessed 14 August 2014; http://en.wikipedia.org/wiki/Pleasureland_Southport accessed 14 August 2014 says it opened in 1912.

44  Anya Chapman and Duncan Light, this volume; also Anya Chapman, 'Coasters at the Coast', *Context* 130 (July 2013), p. 20.

45  Dave Russell, 'Iles, (John) Henry (1871–1951)', *Oxford Dictionary of National Biography* (Oxford: Oxford University Press, 2004); online edn, May 2006: http://www.oxforddnb.com/view/article/48777 accessed 12 April 2013.

46  Evans, *Dreamland Remembered*, pp. 11–14; Morley, *Margate and its Theatres*, p. 137.

47  Kane, 'A Whirl of Wonder!', p. 195; Kane, *The Architecture of Pleasure*, p. 171.

48  Evans, *Dreamland Remembered*, pp. 31–5; Kane, *The Architecture of Pleasure*, p. 175.

49  See Ian Trowell, this volume. Also Deborah Philips, *Fairground Attractions: A Genealogy of the Pleasure Ground* (London: Bloomsbury Academic, 2012), pp. 14–15; NHLE 1374685.

50  http://www.aecom.com/deployedfiles/Internet/Capabilities/Economics/_documents/Theme%20Index%202011.pdf accessed 14 August 2014.

51  John Prest, 'Peel, Sir Robert, Second Baronet (1788–1850)', *Oxford Dictionary of National Biography* (Oxford: Oxford University Press, 2004); online edn, May 2009: http://www.oxforddnb.com/view/article/21764 accessed 14 August 2014; Philips, *Fairground Attractions*, p. 17.

52  http://en.wikipedia.org/wiki/Chessington_World_of_Adventures accessed 14 August 2014.

53  A Cockney, 'A Week's Journal at Margate', *The Mirror* (1823), p. 375; *Morning Chronicle* (3 October 1823), p. 3.

54  Bill Curtis, *Blackpool Tower* (Lavenham: Dalton, 1988), p. 31.

55  Martin Easdown, *Piers of Kent* (Stroud: Tempus Publishing, 2007), p. 110.

56  Easdown, *Piers of Kent*, p. 76.

57  Anthony Wills and Tim Phillips, *British Seaside Piers* (London: English Heritage, 2014), pp. 246–52.

58  Keith Parry, *Resorts of the Lancashire Coast* (Newton Abbot: David and Charles, 1983), pp. 146–52; Preedy, *Roller Coasters*, p. 12.

59  Peter Bennett, *Blackpool Pleasure Beach: A Century of Fun* (Blackpool: Blackpool Pleasure Beach, 1996), pp. 19–21.

60  Kane, 'A Whirl of Wonder!', p. 62; Kane, *The Architecture of Pleasure*, pp. 40, 85.

61  Toulmin, *Blackpool Pleasure Beach*, p. 16.

62  Walton, *Riding on Rainbows*, p. 29; Toulmin, *Blackpool Pleasure Beach*, p. 21.

63  Walton, *Riding on Rainbows*, p. 24.

64  Bennett, *Blackpool Pleasure Beach*, p. 23; Walton, *Riding on Rainbows*, p. 30.

65  Bennett, *Blackpool Pleasure Beach*, p. 31; Walton, *Riding on Rainbows*, p. 33; Toulmin, *Blackpool Pleasure Beach*, p. 25.

66  Kane, 'A Whirl of Wonder!', p. 172; Walton, *Riding on Rainbows*, pp. 47–9; Kane, *The Architecture of Pleasure*, p. 147. Strickler also oversaw the construction of Noah's Arks at Morecambe and Southport in 1930: Toulmin, *Blackpool Pleasure Beach*, p. 33.

67 Bennett, *Blackpool Pleasure Beach*, p. 45; Walton, *Riding on Rainbows*, pp. 47, 50; Toulmin, *Blackpool Pleasure Beach*, p. 33; Kane, *The Architecture of Pleasure*, p. 148.
68 Bennett, *Blackpool Pleasure Beach,* p. 63; Walton, *Riding on Rainbows*, p. 61.
69 Bennett, *Blackpool Pleasure Beach*, p. 65; Bennett, *Roller Coaster*, p. 40; Walton, *Riding on Rainbows*, p. 64; Toulmin, *Blackpool Pleasure Beach*, pp. 43, 127.
70 http://rcdb.com/r.htm?ot=2&pe=6939 accessed 14 August 2014.
71 Bennett, *Blackpool Pleasure Beach*, pp. 58–9; Kane, 'A Whirl of Wonder!', p. 176; Toulmin, *Blackpool Pleasure Beach*, p. 39; Kane, *The Architecture of Pleasure*, pp. 150–51.
72 Alan Powers, 'Emberton, Joseph (1889–1956)', *Oxford Dictionary of National Biography* (Oxford: Oxford University Press, 2004): http://www.oxforddnb.com/view/article/37396 accessed 14 August 2014; Bennett, *Blackpool Pleasure Beach*, pp. 61–2; Toulmin, *Blackpool Pleasure Beach*, p. 33.
73 Walton, *Riding on Rainbows*, pp. 91–2; Toulmin, *Blackpool Pleasure Beach*, p. 82.
74 Bennett, *Blackpool Pleasure Beach*, p. 98; Walton, *Riding on Rainbows*, p. 97.
75 Bennett, *Blackpool Pleasure Beach*, pp. 129ff; Walton, *Riding on Rainbows*, pp. 112–15; Toulmin, *Blackpool Pleasure Beach*, p. 133.
76 Walton, *Riding on Rainbows*, p. 114; Toulmin, *Blackpool Pleasure Beach*, p. 102.
77 Edward Goate 'The Old Switchback and the Hotchkiss Bicycle Railway', *Yarmouth Archaeology* (1994), 17–19; http://en.wikipedia.org/wiki/Great_Yarmouth_Pleasure_Beach accessed 14 August 2014.
78 Historic England Aerofilms Collection EPW001875.
79 http://wicksteedpark.org/rides/ accessed 14 August 2014.
80 NHLE 1390517. The company, now named Wicksteed Leisure Limited, is still a manufacturer of children's playground equipment: http://www.wicksteed.co.uk/ accessed 14 August 2014.
81 Among these is http://rcdb.com/ accessed 14 August 2014, and unpublished research carried out by Nick Laister.
82 See Caroline Ford, this volume.
83 NRHP 78000753: http://rcdb.com/203.htm; NRHP 87000764: http://rcdb.com/204.htm accessed 14 August 2014.
84 NRHP 91000907: http://rcdb.com/9250.htm accessed 14 August 2014.
85 Easdown, *Piers of Kent*, p. 110; Kane, 'A Whirl of Wonder!', p. 199; Evans, *Dreamland Remembered*, pp. 53–4.

# Select bibliography

Adams, J. A., *The American Amusement Park Industry: A History of Technology and Thrills*, Boston: Twayne, 1991.

Aron, C., *Working At Play: A History of Vacations in the United States*, New York: Oxford University Press, 2001.

Ashby, L., *With Amusement for All: A History of American Popular Culture since 1830*, Lexington: University Press of Kentucky, 2006.

Bailey, P., *Leisure and Class in Victorian England*, London: Routledge and Kegan Paul, 1978.

Barker, N., Brodie, A., Dermott, N., Jessop, L. and Winter, G., *Margate's Seaside Heritage*, London: English Heritage, 2007.

Bennett, D., *Roller Coaster*, London: Aurum, 1998.

Bennett, P., *Blackpool Pleasure Beach: A Century of Fun*, Blackpool: Blackpool Pleasure Beach, 1996.

Bennett, T., 'Hegemony, Ideology, Pleasure: Blackpool', in T. Bennett, C. Mercer and J. Woollacott (eds), *Popular Culture and Social Relations*, Milton Keynes: Open University Press, 1986, pp. 135–54.

Bennett, T., *The Birth of the Museum: History, Theory, Politics*, London: Routledge, 1995.

Bennett, T., Mercer, C. and Woollacott, J. (eds), *Popular Culture and Social Relations*, Milton Keynes: Open University Press, 1986.

Boym, S., *The Future of Nostalgia*, New York: Basic Books, 2001.

Bozdoğan, S., *Modernism and Nation Building: Turkish Architectural Culture in the Early Republic*, Seattle: University of Washington Press, 2001.

Bozdoğan S. and Kasaba, R. (eds), *Rethinking Modernity and National Identity in Turkey*, Seattle: University of Washington Press, 1997.

Brodie, A., *Historic Amusement Parks and Fairground Rides: Introductions to Heritage Assets*, London: Historic England, 2015.

Brodie, A. and Winter, G., *England's Seaside Resorts*, London: English Heritage, 2007.

Brodie, A. and Whitfield, M., *Blackpool's Seaside Heritage*, London: English Heritage, 2014.

Bryman, A., *The Disneyization of Society*, London: Sage, 2004.

Cartmell, R., *The Incredible Scream Machine*, Ohio: Amusement Park Books, 1987.

Clavé, S.A., *The Global Theme Park Industry*, Wallingford: CABI, 2007.

Copnall, S., *Pleasureland Memories: A History of Southport's Amusement Park*, St Albans: Skelter Publishing, 2005.

Cross, G., 'Crowds and Leisure', *Journal of Social History*, 39, 3 (Spring 2006), pp. 631–50.

Cross, G. and Walton, J. K., *The Playful Crowd: Pleasure Places in the Twentieth Century*, New York: Columbia University Press, 2005.

Crowe, K., *Kursaal Memories: A History of Southend's Amusement Park*, St Albans: Skelter Publishing, 2003.

Cunningham, H., *Leisure in the Industrial Revolution*, New York: St Martin's Press, 1980.

Davis, S., *Spectacular Nature: Corporate Culture and the Sea World Experience*, Berkeley: University of California Press, 1997.

Denson, C., *Coney Island Lost and Found*, Berkeley: Ten Speed Press, 2002.

Desmond, J., *Staging Tourism: Bodies on Display from Waikiki to Sea World*, Chicago: University of Chicago Press, 1999.

Easdown, M., *Amusement Park Rides*, Oxford: Shire Publications, 2012.

Easdown, M. and Sage, L., *The Demise of the Rotunda*, Folkestone: Marlinova, 2008.

Evans, N., *Dreamland Revived: The Story of Margate's Famous Amusement Park*, rev. edn, Whitstable: Bygone Publishing, 2014.

Ford, C., 'The First Wave: The Making of a Beach Culture in Sydney, 1810–1920', unpublished Doctoral thesis, University of Sydney, 2007.

Ford, C., 'A Summer Fling: The Rise and Fall of Aquariums and Fun Parks on Sydney's Ocean Coast 1885–1920', *Journal of Tourism History*, 1, 2 (2009), pp. 95–112.

Ford, C., *Sydney Beaches: A History*, Sydney: New South Books, 2014.

Funnell, C. E., *By the Beautiful Sea: The Rise and High Times of That Great American Resort, Atlantic City*, New York: Knopf, 1975.

Gilmore, J. H. and Pine, J., *Authenticity: What Consumers Really Want*, Boston: Harvard Business School Press, 2007.

Gray, F., *Designing the Seaside: Architecture, Society and Nature*, London: Reaktion, 2006.

Harrison, R., 'What is Heritage?', in R. Harrison (ed.), *Understanding the Politics of Heritage*, Manchester: Manchester University Press/Open University, 2010, pp. 5–42.

Harrison, R., *Heritage: Critical Approaches*, London: Routledge, 2013.

Hawkey, A., *The Amazing Hiram Maxim: An Intimate Biography*, Staplehurst: Spellmount, 2001.

Hewison, R., *The Heritage Industry: Britain in a Climate of Decline*, London: Methuen, 1987.

Kahrl, A., *The Land Was Ours: African American Beaches from Jim Crow to the Sunbelt South*, Cambridge, Massachusetts: Harvard University Press, 2012.

Kane, J., 'A Whirl of Wonder!' British Amusement Parks and the Architecture of Pleasure 1900–1939', unpublished PhD thesis, University of London, 2007.

Kane, J., *The Architecture of Pleasure: British Amusement Parks 1900–1939*, Farnham: Ashgate, 2013.

Kasson, J. F., *Amusing the Million: Coney Island at the Turn of the Century*, New York: Hill & Twang, 1978.

King, M. J., 'The Theme Park: Aspects of Experience in a Four-Dimensional Landscape', *Material Culture*, 34 (2002), pp. 1–15.

Kyriazi, G., *The Great American Amusement Parks: A Pictorial History*, Secaucus, New Jersey: Citadel Press, 1976.

Laister, N. and Page, D., 'Request for Spot Listing of the Scenic Railway Roller Coaster, Dreamland Amusement Park, Kent', Wantage: Nick Laister, 2001.

Liebowitz, S., *Steel Pier, Atlantic City: Showplace of the Nation*, West Creek, New Jersey: Down the Shore Publishing, 2009.

Lowenthal, D., *The Heritage Crusade and the Spoils of History*, Cambridge: Cambridge University Press, 1998.

Lowenthal, D., 'The Past as a Theme Park', in T. Young and R. Riley (eds), *Theme Park Landscapes: Antecedents and Variations*, Washington: Dumbarton Oaks, 2002, pp. 11–23.

Lukas, S. A., *Theme Park*, London: Reaktion, 2008.

Macdonald, G. and Alsford, S., 'Museum and Theme Parks: Worlds in Collision?' *Museum Management and Curatorship*, 14, 2 (1995), pp. 129–47.

Misa, T., Brey, P. and Feenberg, A (eds), *Modernity and Technology*, Cambridge, Massachusetts/London: MIT Press, 2003.

Mohun, A., 'Design for Thrills and Safety: Amusement Parks and the Commodification of Risk, 1880–1929', *Journal of Design History*, 14, 4 (2001), pp. 291–306.

Nasaw, D., *Going Out: The Rise and Fall of Public Amusements*, New York: Basic Books, 1993.

Neil, J. M., 'The Roller Coaster: Architectural Symbol and Sign', *Journal of Popular Culture*, 15, 1 (1981), pp. 108–15.

Nelson, S., 'Walt Disney's EPCOT and the World's Fair Performance Tradition', *The Drama Review*, 30, 4 (1986), pp. 106–46.

Nye, D. E., *American Technological Sublime*, Cambridge, Massachusetts/London: MIT Press, 1994.

Parry, K., *Resorts of the Lancashire Coast*, Newton Abbot: David and Charles, 1983.

Philips, D., *Fairground Attractions: A Genealogy of the Pleasure Ground*, London: Bloomsbury Academic, 2012.

Pilat, O. and Ranson, J., *Sodom by the Sea, An Affectionate History of Coney Island*, Garden City, New York: Doubleday, 1941.

Preedy, R. E., *Roller Coasters: Their Amazing History*, Leeds: Robert E. Preedy, 1992.

Preedy, R. E, *Roller Coaster: Shake, Rattle and Roll!*, Leeds: Robert E. Preedy, 1996.

Register, W., *The Kid of Coney Island: Fred Thompson and the Rise of American Amusements*, New York: Oxford University Press, 2001.

Relph, E., *Place and Placelessness*, London: Pion Press, 1976.

Ritzer, G., *The McDonaldization of Society: An Investigation Into the Changing Character of Contemporary Social Life*, London: Pine Gorge, 1996.

Rojek, C., *Ways of Escape: Modern Transformations in Leisure and Travel*, Basingstoke/London: Macmillan Press, 1993.

Rutherford, S., *The American Roller Coaster*, Osceola, Wisconsin: MBI Publishing, 2000.

Shields, R., *Places on the Margin: Alternative Geographies of Modernity*, London: Routledge, 1991.

Simon, B., *Boardwalk of Dreams: Atlantic City and Fate of Urban America*, New York: Oxford University Press, 2004.

Smith, L., *Uses of Heritage*, London: Routledge, 2006.

Smith, L. and Waterton, E., '"The Envy of the World?": Intangible Heritage in England', in L. Smith and N. Akagawa (eds), *Intangible Heritage*, London: Routledge, 2009, pp. 289–302.

Starsmore, I., *English Fairs*, London: Thames and Hudson, 1975.

Stewart, S., *On Longing: Narrative of the Miniature, the Gigantic, the Souvenir, the Collection*, Baltimore: John Hopkins University Press, 1984.

Toulmin, V., *Blackpool Pleasure Beach: More Than Just an Amusement Park*, Hathersage: Boco, 2011.

Urbanowicz, S. J., *The Roller Coaster Lover's Companion: A Thrill Seeker's Guide to the World's Best Coasters*, rev. edn, New York: Citadel Press, 2002.

Uzzell, D., 'Heritage that Hurts: Interpretation in a Post-modern World', in D. Uzzell and R. Ballantyne (eds), *Contemporary Issues in Heritage and Environmental Interpretation: Problems and Prospects*, London: The Stationary Office, 1998, pp. 152–71.

Walton, J. K., 'The Demand for Working-class Seaside Holidays in Victorian England', *Economic History Review*, 34, 2 (1981), pp. 249–65.

Walton, J. K., *Wonderlands by the Waves: A History of the Seaside Resorts of Lancashire*, Preston: Lancashire County Books, 1992.

Walton, J. K., *Blackpool*, Edinburgh: Edinburgh University Press, 1998.

Walton, J. K., *The British Seaside: Holidays and Resorts in the Twentieth Century*, Manchester: Manchester University Press, 2000.

Walton, J. K., 'Popular Playgrounds: Blackpool and Coney Island, c. 1880–1970', *Manchester Region History Review*, 17, 1 (2004), pp. 51–61.

Walton, J. K., *Riding on Rainbows: Blackpool Pleasure Beach and its Place in British Popular Culture*, St Albans: Skelter Publishing, 2007.

Walton, J. K. and Wood, J., 'World Heritage Seaside', *British Archaeology* 90 (2006), pp. 10–15.

Walton, J. K. and Wood, J., 'History, Heritage and Regeneration of the Recent Past: The British Context', in N. Silberman and C. Liuzza (eds), *Interpreting The Past V, Part 1. The Future of Heritage: Changing Visions, Attitudes and Contexts in the 21st Century*, Brussels: Province of East-Flanders, Flemish Heritage Institute and Ename Center for Public Archaeology and Heritage Presentation (2007), pp. 99–110.

Walton, J. K. and Wood, J., 'Reputation and Regeneration: History and the Heritage of the Recent Past in the Re-making of Blackpool', in L. Gibson and J. Pendlebury (eds), *Valuing Historic Environments*, Farnham: Ashgate, 2009, pp. 115–37.

Watson, S., 'History Museums, Community Identities and Sense of Place', in S. J. Knell, S. MacLeod and S. Watson (eds), *Museum Revolutions: How Museums Change and Are Changing*, London: Routledge, 2007, 160–72.

Webb, D., 'Bakhtin at the Seaside: Utopia, Modernity and the Carnivalesque', *Theory, Culture and Society*, 22, 3 (2005), pp. 121–38.

Weinstein, R. M., 'Disneyland and Coney Island: Reflections on the Evolution of the Modern Amusement Park', *Journal of Popular Culture*, 26, 1 (1992), pp. 131–64.

Wilkes, P., *Fairground Heritage*, Burton-upon-Trent: Trent Valley Publications, 1989.

Williamson, R. W., *Popular Recreations in English Society 1700–1850*, Cambridge: Cambridge University Press, 1973.

Wills, A. and Phillips, T., *British Seaside Piers*, London: English Heritage, 2014.

Wolcott, V., *Race, Riots, and Roller Coasters: The Struggle Over Segregated Recreation in America*, Philadelphia: University of Pennsylvania Press, 2012.

Wood, J., 'From Port to Resort: Art, Heritage and Identity in the Regeneration of Margate', in P. Borsay and J. K. Walton (eds), *Resorts and Ports: European Seaside Towns Since 1700*, Bristol: Channel View Publications, 2011, pp. 197–218.

Wyatt, M., *White Knuckle Ride*, London: Salamander, 1996.

Zukin, S., *Landscapes of Power: From Detroit to Disney World*, Los Angeles: University of California Press, 1993.

# Index

# Taylor & Francis eBooks

## Helping you to choose the right eBooks for your Library

Add Routledge titles to your library's digital collection today. Taylor and Francis ebooks contains over 50,000 titles in the Humanities, Social Sciences, Behavioural Sciences, Built Environment and Law.

**Choose from a range of subject packages or create your own!**

**Benefits for you**

» Free MARC records
» COUNTER-compliant usage statistics
» Flexible purchase and pricing options
» All titles DRM-free.

**Benefits for your user**

» Off-site, anytime access via Athens or referring URL
» Print or copy pages or chapters
» Full content search
» Bookmark, highlight and annotate text
» Access to thousands of pages of quality research at the click of a button.

## eCollections – Choose from over 30 subject eCollections, including:

| | |
|---|---|
| Archaeology | Language Learning |
| Architecture | Law |
| Asian Studies | Literature |
| Business & Management | Media & Communication |
| Classical Studies | Middle East Studies |
| Construction | Music |
| Creative & Media Arts | Philosophy |
| Criminology & Criminal Justice | Planning |
| Economics | Politics |
| Education | Psychology & Mental Health |
| Energy | Religion |
| Engineering | Security |
| English Language & Linguistics | Social Work |
| Environment & Sustainability | Sociology |
| Geography | Sport |
| Health Studies | Theatre & Performance |
| History | Tourism, Hospitality & Events |

For more information, pricing enquiries or to order a free trial, please contact your local sales team:
www.tandfebooks.com/page/sales

Printed and bound by CPI Group (UK) Ltd, Croydon, CR0 4YY

29/10/2024

01780523-0002